John Scandrett Harford

Recollections of William Wilberforce, Esq. M.P.

For the County of York During nearly thirty years - with brief notices of some of his personal friends and contemporaries. Second Edition

John Scandrett Harford

Recollections of William Wilberforce, Esq. M.P.
For the County of York during nearly thirty years - with brief notices of some of his personal friends and contemporaries. Second Edition

ISBN/EAN: 9783337776206

Printed in Europe, USA, Canada, Australia, Japan

Cover: Foto ©berggeist007 / pixelio.de

More available books at **www.hansebooks.com**

WILLIAM WILBERFORCE.

RECOLLECTIONS

OF

WILLIAM WILBERFORCE, ESQ.

M.P. FOR THE COUNTY OF YORK

DURING NEARLY THIRTY YEARS:

WITH BRIEF NOTICES OF SOME OF

HIS PERSONAL FRIENDS AND CONTEMPORARIES.

BY

JOHN S. HARFORD, Esq. D.C.L. F.R.S.

SECOND EDITION.

LONDON:
LONGMAN, GREEN, LONGMAN, ROBERTS, & GREEN.
1865.

PREFACE.

THE NAME of WILBERFORCE is honoured and revered throughout the civilised world as that of a Christian philanthropist, and in particular as the principal achiever of that great work of humanity, the Abolition of the Slave Trade. His eminent talents as an orator and a statesman, and his admirable qualities both of head and heart, procured for him also an influence for good, both in and out of Parliament, and among men of all parties, scarcely equalled by any of his contemporaries. This influence throughout the whole of his career was studiously exercised to promote the glory of God and the best interests of his fellow-creatures.

His sons, in their biography of this great and good man, have ably illustrated these particulars. They have also opened to the public the interior of their father's religious and social life by numerous extracts from his Diary and Correspondence; but we think it may not be displeasing to the reader to listen to the testimony, to the same effect, of one who was wholly unconnected with him by family ties, but who was honoured with his intimate friendship throughout a period of upwards of twenty-one years of his eventful life.

Although the illustration of his private qualities is the chief object of development in these pages, they are interspersed with recollections of his public life, and with anecdotes of many of the distinguished men of his time with whom he was familiar. It is hoped that the additional authentic data here presented, and collected chiefly from his own lips, may add to the veneration and esteem with which his name and character are regarded. They will also give some slight idea of him as he was seen in the retirement of the country, among a select circle of friends, whom he daily delighted by his wisdom and wit, by the charm of his colloquial powers, and by the attractive influence of his religious and benevolent affections.

Various facts and anecdotes, and some important passages illustrative of his early life and of a change in his religious opinions occurring at that time, which may be found almost verbatim in the Biography, were gleaned, with the present writer's consent, from the manuscripts which form the staple of the present volume, but the greater part has not till now seen the light.

In confirmation of this statement the following extracts from letters addressed to the writer by the late Rev. Robert I. Wilberforce, will prove he was not slow to acknowledge the aid which it was in the present writer's power thus to render.

'EAST FARLEIGH: *April* 25, 1838.

'My dear Friend,—At length I trust that with this letter you will receive a copy of my father's Life, of which my brother and myself beg your acceptance. You will see how greatly your aid has contributed to it.'

And in earlier letters:—

'I have made great use of the extracts which I made myself from your papers when I was at Blaise Castle.' 'I shall very soon arrive at the place when I can make use of any records of conversation of a miscellaneous kind, such as several exquisite passages you showed me.' 'I have some remarks which I copied myself out of your papers, about Castlereagh, Pitt, Sheridan, Fox, &c.; also about English villas, Dr. Johnson's sentiments on Sacred Poetry, &c. &c. I mean to introduce them as specimens of conversation.'

The writer himself, it will be found, is indebted to the Biography for some valuable facts and memoranda, particularly for the account of the closing scenes of Mr. Wilberforce's life.

The writer's object has been throughout to avoid reference to himself, as far as the course of the narrative would permit, and to make Mr. Wilberforce speak for himself.

CONTENTS.

CHAPTER I.
1812.

Author's Introduction to Mr. Wilberforce—His First Impressions—Mr. W.'s Domestic Habits—West, Garrick, and Nollekens—Death of Mr. Perceval—Author's Visit to Ireland—Visitors at Bellevue . . . . PAGE 1

CHAPTER II.
1813.

Renewal of the Charter of the East India Company; involving the Question, What Degree of Support to the Church of England in that Country, and to Missionaries, should be conceded by the Legislature?—Animated Debates on the Subject, in which Mr. W. took a prominent part 25

CHAPTER III.
1813 TO 1815.

Mr. W.'s Delight in the Beauties of Nature—His Hints as to the Best Modes of preparing for Parliamentary Speaking—Madame de Staël seeks his Acquaintance—Retreat of the French Army from Moscow, and great Political Events which followed, ending in the Peace of 1814—Disappointment of Mr. W. at the Want of Securities in that Treaty for the Abolition of the Slave Trade—His striking Appeal in allusion to it—Lord Castlereagh's Reply—Presence of Crowned Heads and Illustrious Foreigners in London at this Epoch—Mr. W. waits on the Emperor Alexander by his express Command—Public Meeting at Freemasons' Tavern with respect to the Defects of the Treaty of Peace as to the Slave Trade—Splendid Illuminations in London—Mr. W. visits them—Duke of Wellington a cordial Abolitionist—Death of Mr. Thornton and of various Friends. . . . . . . PAGE 42

CHAPTER IV.
1815 TO 1818.

Pleasant Excursions and Visits in Company with Mr. Wilberforce—Author's Residence in Rome—His Acquaintance with Cardinal Consalvi, the Papal Secretary of State—Mr. W. requests the Author to place before the Cardinal important Documents relative to the Slave Trade—The Cardinal arranges for him in consequence a Private Interview with the Pope, Pius VII.—Correspondence with Mr. W. 67

CHAPTER V.
1817 TO 1820.

Author's Return to England—Renewed Intercourse with Mr. W.—Visit with him to Newgate, and Readings of Mrs. Fry to the Female Prisoners—Visit from Mr. and Mrs. W. and Family—Slaves sold from England into Ireland—Meetings with Mr. W. in London, and in the Country—Walks with him at Bath, and Anecdotes related by him of Lord North and others—The late Lord Teignmouth, and Tribute to his Memory—Mr. W.'s Bachelor Life in Palace Yard—His Estimate of Robert Hall as a Preacher—Visit from Mr. W. in December 1820—Various Anecdotes—Mr. W.'s Work, entitled 'A Practical View of Christianity;' Remarks on, and various Extracts from it 84

CHAPTER VI.
1821 TO 1824.

Various Meetings with Mr. W. at Kensington Gore and elsewhere—Dinner to meet Sir Walter Scott at Clapham at Sir Robert Inglis's—Change of Residence from Kensington Gore to Marden Park—Mode of Life there—Mr. W. under Domestic Affliction at the Close of 1821—Christian Happiness in Death in the Case of Lady ——. —Stay at Marden Park in 1822—Baron de Staël— Mr. W.'s Visit to Norfolk: Acquaintance and Friendship with the Gurney Family—Mr. Buxton's Speech on Slavery—Death and Character of Mr. Charles Grant—Bishop Heber's Departure for India—Mr. W.'s Thoughts as to retiring from Public Life, 1824—His temporary Residence near Uxbridge—The Author revisits Ireland and Bellevue—Stays with Mr. W. at Bath; and receives him and Mrs. W. for about a Fortnight at his own Residence. . . . . PAGE 111

CHAPTER VII.

How Mr. W. was first induced to take up the Slave-trade Question —Granville Sharp—While staying at Holwood, Mr. Pitt urges him to bring it forward—He opens it in the House in 1789— Determined Opposition—Professor Sedgwick's Recollections of the brilliant Debates it called forth—Pitt's memorable Speech— Windham's Remarks on it—Cowper's Sonnet to Mr. W.— Mr. W.'s Recollections of Mr. Pitt—They visit France in 1783 accompanied by Mr. Eliot—Amusing Incidents—Mr. Pitt, Premier— Mr. W. becomes M.P. for the County of York—Pitt, Fox, Canning, Lord Harrowby 138

CHAPTER VIII.

1825 TO 1828.

Retirement of Mr. W. from Parliament—The Author visits Staffa and Iona—Honourable Mention of Mr. W. there—A Visit from Mr. W. in December 1825—His Remarks on Castlereagh— Sheridan—Dr. Johnson—Burke—Various Anecdotes of Distinguished Men—Dr. Carey, the Baptist Missionary—Eloquence and Wit of Canning—Lord Brougham—A Visit to Mr. W. in Bath in 1827—Mr. W. visits Yorkshire—Attentions paid him there—In 1828 Visits to him at Bath, &c.—Mr. W.'s 70th Birthday . 158

CHAPTER IX.

Mr. W. details some of the leading Particulars of his Life . 196

CHAPTER X.

1830 TO 1833.

Letters from Mr. W.—A Visit from him in November—He quits Highwood Hill in 1831—Visits to him at Bath—The Reform Crisis—Bristol Riots during a Visit from Mr. W.—In 1832 Domestic Affliction—Visit to Mr. Stephen—Visit from Mr. and Mrs. W. and Bishop Ryder—This was Mr. W.'s last Visit—Vigorous Letter from him in February 1833—His Speech at Maidstone in 'April—Letter from 'him at Bath in May—His Interview with Mr. J. J. Gurney—Leaves Bath for London in July—Illness and Death—His Funeral—And Last Honours paid to him . . . . PAGE 223

CHAPTER XI.

Incidental Remarks on his Character and Domestic Habits—Portrait by Lawrence, and ditto by Richmond—His Statue in Westminster Abbey—Letter from Provost of Oriel 255

CHAPTER XII.

Brief Recollections of Mrs. Hannah More . . . 268

CHAPTER XIII.

Sketch of the Life and Character of the Rev. R. C. Whalley, Rector of Chelwood 294

POSTSCRIPT 323

Erratum.

Page 142, line 18 from top, *for* limine *read* lumine.

Harford's 'Wilberforce,' 2nd Edit.

RECOLLECTIONS

OF

WILLIAM WILBERFORCE, ESQ.

CHAPTER I.

1812.

Author's Introduction to Mr. Wilberforce—His First Impressions—Mr. W.'s Domestic Habits—West, Garrick, and Nollekens—Death of Mr. Perceval—Author's Visit to Ireland—Visitors at Bellevue.

IN the year 1812 I had the pleasure of becoming acquainted with Mr. Wilberforce. I had long and largely partaken of the affectionate veneration generally felt for his name and character as an eminent Christian statesman, and as the undaunted champion, throughout numerous years of labour, obloquy, and trial, of the abolition of the African Slave Trade.

It was, therefore, no small gratification to me to be invited by Mr. Henry Thornton, his distinguished friend and coadjutor in all his noble efforts to do good both in and out of Parliament, to dine with him at his house in Palace Yard on an occasion when

Mr. Wilberforce was expected. In the morning of that day I had heard him speak, for the first time, on behalf of the British and Foreign Bible Society, at Freemasons Hall. The fine tones of his voice, and his winning address, imparted the fullest possible effect to the noble and benevolent sentiments which he poured forth on this occasion, and which were received with enthusiasm. One sentence fastened on my memory: 'The Trophies of our Society are bought with no tears, excepting tears of gratitude and joy.'

Soon after the guests had assembled at the dinner-table he made his appearance. I seem at this moment to see him coming in with a smiling animated countenance, and a lively vivacity of movement and manner, exchanging kind salutations with his friends, whose faces lighted up with pleasure at his entry. I was much struck throughout the evening by the ease of his manners, the playful brilliancy of his thoughts, and his unaffected kindness to all around him. He was then fifty-two. Those who never saw him, excepting in his later years, when his head almost reclined on his chest, from the effect of age on an extremely delicate frame, cannot form a just idea of him at the time now referred to. Though always somewhat of a valetudinarian, the spring and elasticity of his movements at this period, the comparative erectness of his figure, and the glow upon his cheek, formed a strong contrast to the bodily infirmities which gradually stole upon him in advancing life. His frame was at all times extremely spare, and seemed to indicate that the ethereal inhabitant within was burdened with as little as possible of corporeal adjuncts: but from this form so slight proceeded a voice of uncommon compass and richness, whose varying and impressive tones, even in common conversation, bespoke the powers of the orator.

His eyes, though small, beamed with the expression of acute intelligence, and of comprehension quick as lightning, blended with cordial kindness and warmth of heart. A peculiar sweetness and playfulness characterised his mouth, his forehead was ample, and his head well formed; and though there was not a single handsome feature, yet the mingled emanations of imagination and intellect, of benevolence and vivacity, diffused over his countenance a sunny radiance which irresistibly attracted the hearts of all who approached him. An eye-glass, of which he made constant use, and a diamond brooch, were appendages of his person. At this time, and till within a very few years of his death, he wore powder, and his dress and appearance were those of a thorough gentleman of the old school.

I had brought to town with me a letter of introduction to him from his old and much-valued friend Hannah More, which I took an early opportunity of leaving at his residence at Kensington Gore. It quickly procured me an invitation to dinner, which I subjoin because it is characteristic of himself; it was followed by many more; and such was the confidence with which his cordial and friendly manner inspired me, that I found myself quite at ease with him, and was able to ask his opinion upon any topics, either of private or of public interest, with entire freedom.

'Kensington Gore: Saturday Evening, May 9, 1812.

'My dear Sir,—Ever since I had the pleasure of seeing you on Wednesday last I have been so incessantly occupied that I really forgot to execute the intention I had formed of endeavouring to effect a meeting with you, in however hurried a way, which might give us mutually a right to claim each other's acquaintance when an opportunity of quiet intercourse should occur. Almost the only way in which I can see my friends during the sitting of Parliament is sometimes at a mutton-chop dinner before the House of Commons—to which I go over when business begins. Will you fix any day for favouring me with your company in that way at the New Palace Yard Hotel? But though I never see *company* (so to be called) on a Sunday, yet I sometimes ask a friend or two, who likes our way of going on, upon that day. If you stay here on any Sunday—to-morrow, or any other —I shall be happy to see you. We dine at three, and go to the Lock Chapel at six; and both before dinner and supper you would have the command of your own time. I can promise you a well-aired bed and a cordial reception. I am, always with esteem and regard, my dear Sir,

'Your faithful Servant,
(Signed) 'W. WILBERFORCE.
'—— Harford, Esq.'

Before I left London he put me upon the complete footing of a friend, by conducting me himself to what he called his bachelor's room, and by pressing me to occupy it whenever I felt inclined, without previous notice or ceremony, and he made me feel that these were not mere words of compliment by often chiding me in a pleasant way for not availing myself more frequently of this privilege.

Breakfast was the meal at which Mr. Wilberforce's friends could be sure of meeting him during the sitting of Parliament. He seldom gave dinners except on Saturdays, and indeed did not often get back to Kensington Gore till the evening. Sometimes, when the House had sat very late, he slept in London. The nocturnal vigils of Parliament had rendered his breakfast-hour late, and it was still further delayed by his undeviating practice of never leaving his dressing-room without securing time for devotional exercises and for reading the Bible. How

much he felt on this subject will appear by the following extract from a letter to one of his sons :*—
'O my dearest boy, let me earnestly conjure you not to be seduced into neglecting, curtailing, or hurrying over your morning prayers. Of all things, guard against neglecting God in the closet. There is nothing more fatal to the life and power of religion—nothing which makes God more certainly withdraw His grace' (March 1815).

When he made his appearance it was the signal for family prayer. He commenced with a portion of Scripture, making a comment upon it of a practical and devotional tendency, marked throughout by simple and familiar illustration, adapted to the capacities of his servants. He ended with a prayer, which, though he always had a book on such occasions open before him, was evidently oftener than not the effusion of his own winged soul.

He frequently spoke of the refreshment which he felt from the religious observance of the Lord's Day. I was often his guest from Saturday till Monday, and shall ever retain a delightful impression of the happiness of a Sunday spent in his society. There was nothing of affected seriousness in his deportment or manner on this day. All was easy, natural, and cheerful. It was obvious that he regarded not the

* Life, vol. iv. p. 248.

obligations of its sacred duties as a burdensome form, but as a high and exalted privilege—a source of efficacious improvement in holy things. It was his constant habit to attend church, both in the morning and evening. The devotion and concentration of his manner when there were truly impressive. He had left the world and its cares behind him, and had ascended the Mount to contemplate, to hold converse with, and to adore his God and Saviour. On returning from morning church, he usually spent a short time in the society of his family and friends, and in the fine season of the year occasionally strolled with them in the garden, when his conversation caught a sacred influence from the employments and objects of the day. The dinner-hour was about three, to admit of going to evening church, and it was his invariable habit to seek the retirement of his private room for at least an hour and a half before that period for devotional purposes. I almost seem, at this moment, to behold him on one of these occasions passing through the ante-room of his library, when about thus to retire, with a folio under his arm, and stopping me with a smile to tell me that his companion was a volume of Baxter's Practical Works. Baxter was one of his favourite divines, and he shared this preference with Dr. Johnson, who, as Boswell tells us, on being asked what works of Baxter he would most recommend, replied, 'Sir, all that he has written is good.' During the long days at the close of spring he sometimes went into the garden at a late hour to listen to the nightingales which then abounded at Kensington. None but a few select friends who liked to spend their time as he did were invited to his house on Sundays. In the evenings he sometimes read aloud fine passages from his favourite poet, Cowper, imparting to them the fullest effect by the tones of his musical voice, and he was always ready for conversation at once elevating and instructive.

Thus, even amidst the exciting scenes of the metropolis, and the incessant and anxious occupations of public life, he habitually 'walked with God.' And what was the consequence? While so many other statesmen, acting on the common motives of human ambition, were careworn, disappointed, and ill at ease, he was serene, cheerful, and happy, and at the age of fifty-two had the gaiety and the spirits of a young man of twenty. Let the practical results of true and undefiled religion be appealed to as the test of its excellence, and it will be found that notwithstanding the self-denial, the conflicts, and the trials which an honest and upright adherence to its principles may involve, it is the only source of anything that deserves to be called happiness even on this side the grave.

The guests at Mr. Wilberforce's breakfast-table

were often numerous, consisting of friends who casually dropped in, or of active and benevolent individuals who were promoters of useful or important projects in Parliament, or in a more private way, and who came to him for advice, or assistance, or both—of literary or scientific men—and of some who took an active part in the abolition of the slave trade—and often of clergymen of zeal and superior excellence in their profession. Arthur Young, the agriculturist, was at this time a frequent guest. He was blind, and there was a humility in his manners which, united to the respect and sympathy excited by his venerable appearance and loss of sight, rendered him an object of much interest. In earlier life he had been quite thoughtless on the subject of religion, but the preaching of the Rev. Mr. Cecil and the perusal of Mr. Wilberforce's work on practical Christianity had opened to him new and just views of the nature and object of religion, and he was now, and for years had been, a humble and sincere believer in Jesus Christ.

I was present at a pleasant dinner-party at Mr. Hart Davis's, then M.P. for Bristol, in the spring of 1812, at which Mr. Wilberforce and Mr. West, the venerable President of the Royal Academy, were among the guests. In the course of the evening the two former engaged in an animated conversation upon painting, which finally diverged to the cartoons of

Raphael, of which Mr. West spoke with enthusiastic admiration as the noblest compositions of that great painter. They were not only characterised, he observed, by grandeur of conception and by creative fancy, but were so replete with the truth of nature and with a majestic simplicity, that the impressions of individuals who viewed them for the first time, and who expected in the place of nature to see a hero in every figure, were not seldom those of disappointment and indifference. 'I remember,' he added, 'that Garrick said to me one day, "Well, Mr. West, I have at length visited the cartoons, about which I have heard so much, and to tell you the honest truth, I have been much disappointed." "Indeed," said I, "pray explain yourself." "Why," rejoined he, "there's the figure of Elymas the Sorcerer, for instance: what a shabby-looking old fellow he is, and how strange his attitude." "I should very much like to know, Mr. Garrick," I replied, "if you had to personate Elymas how you would act the part: let me entreat you to indulge the company in this way." After a little solicitation he complied, and going out at a folding door in the apartment quickly returned in the character of Elymas. He had scarcely entered the room, every eye being intently fixed upon him, when I exclaimed, "Mr. Garrick, I entreat you to stand still in your present position without altering a muscle." He did so, when I challenged everyone at all acquainted with the cartoons to say whether it was possible to imagine a closer approach to Raphael's conception of the attitude befitting Elymas on his sudden seizure with blindness than the figure before us. Garrick, in fact, in his just impression of the character he was to personate, had unconsciously put himself into the very attitude of Elymas. His head—half held back, half protruded—and his bent body were protected by his outstretched arms and hands, which cautiously explored his way, while the shoulders a little drawn backwards balanced his position. A more complete practical defence of the painter, or refutation of his own criticism, could not be imagined. Modern sculpture becoming afterwards a topic of conversation, and the statue of Mr. Pitt by Nollekens, in the Senate House at Cambridge, being mentioned, West told a remarkable anecdote connected with it. The managing committee for the statue applied to Canova to undertake the work. That great artist was highly flattered by the compliment, but replied with his characteristic generosity of spirit that it would be presumption in him to accept the commission while London could boast such a sculptor as Nollekens. He was in consequence selected.* It was singular that Nollekens, who had lived so many years in London, and who had

* A Cambridge Professor maintains that Nollekens was selected by open competition, and not by Canova's influence.

executed the busts of so many of the great men of the age, had never seen Mr. Pitt, and was, therefore, unable to form any accurate conception of his person and manner. Various portraits, and a postmortem mask, of his features, were all that he had to assist him. In this difficulty he applied to West, who, from the intimacy with which he had been honoured by George III., had often seen and conversed with the great statesman. He begged him to make a drawing of his personal figure suited to the intended statue. 'I wish, Mr. Nollekens,' replied West, 'you could behold him in your mind's eye exactly as I now see him in mine, on occasion of a remarkable incident. It occurred when the King's mental malady was at its height, and naturally formed the subject of Pitt's deepest anxiety. Circumstances had occurred which rendered it of great consequence that he should himself see the King, and form his own judgment of the state of his mind; but, though he more than once went to Windsor for this purpose, he returned to town on each occasion thwarted by a cabal within the walls of the Castle. At length he went with a firm resolution of effecting, if possible, his purpose in his own way. He came to the door leading into the Royal apartments, and challenged the sentry to let him pass. The man refused, saying his orders were peremptory to let no one pass. Mr. Pitt warned him in so commanding a tone and manner not to impede the Prime Minister that the man yielded; and Pitt, taking the key from the outside, locked the door on the inside, and then walked forward to the suite of rooms occupied by the King. He saw His Majesty, conversed with him, and thus satisfied himself on various points of much interest connected with his malady. 'On quitting the Castle and descending the hill towards Windsor, I happened,' said West, 'to be ascending it. I can never lose the impression of his manner and aspect. He approached me in an attitude of triumph—his head was a little tossed back—he looked like a demi-god. He told me of the part he had just been acting. " And now, Nollekens, I shall endeavour, in the drawing which you request of me, to introduce, if possible, something of the attitude and expression which so much struck me." I did so; and anyone who looks at the statue, bearing in mind this incident, will be able to trace the connection between them.' The conversation embraced various topics connected with the fine arts, and it amused me not a little, in this early period of my acquaintance with Mr. Wilberforce, to witness what I often subsequently had occasion to notice—his impatience of being chained down too long to any one topic. He was himself all point and brevity in relating an anecdote; and as the venerable President was once or twice tediously minute in his descriptions, he fidgeted about on his chair—then got down on one knee looking earnestly at him—then rose, with some exclamation of surprise at what he was listening to, and again quickly changed his position—though it was obvious that he was, on the whole, much interested in the conversation. There had been a little contest between them on going down to dinner, each wishing to yield precedence to the other; but Mr. Wilberforce carried his point, exclaiming, 'Oh, no! I cannot think of going before the President of the Royal Academy.' It was impossible to be in West's company without admiring his countenance and his calm and gentlemanly manner. His features were cast in a fine mould, and were lighted up with the expression of benignity and intellect. He was the Nestor of painters.

He dealt much in anecdotes of the olden time, and delighted to go back, in imagination, to the scenes and acquaintances connected with his opening career of fame in England. Of the great painters of Italy, and of his residence in that land of taste and genius, he talked with fond enthusiasm, and though it was obvious that in comparing himself with them he over-rated his own powers, yet his self-esteem was not obtrusive nor importunate. On one occasion his genius rose into the sublime—for this epithet may justly be applied to the finished oil-sketch for his picture of Death on the Pale Horse. The picture which he made from it widely departed from the sketch, and was a failure. He was fond of relating how much Napoleon had been delighted with this sketch, when it was exhibited at Paris after the Peace of Amiens; and how anxious he had been to purchase it, and how he (West) had declined the honour—telling him that it was expressly painted for His Majesty the King of England, as a model for a large window of stained glass proposed to be put up in St. George's Chapel, at Windsor. Napoleon not only gave him free access to the galleries of the Louvre, but made an appointment to receive him there, on which occasion West mentioned, as a mark of his courteous discrimination, that while they were in the picture gallery Napoleon deferred much to his remarks and criticisms; but the moment they were among the statues, he himself took the lead in commenting upon them.

I will just add one more of his anecdotes, which deserves to be recorded. At one of the exhibitions at Somerset House his attention was forcibly caught by a composition in plaster in the sculpture-room, the subject of which was wild bulls tearing about in furious and animated action. It bore upon it so strongly the impress of superior genius that he was induced to make a memorandum of the name and address of the artist. Some time after the exhibition had closed, this remarkable composition recurred to him, and, feeling anxious to know whether it had sold, he called upon the artist. The young man was much flattered by this mark of approbation and kindness from such a visitor, but told him he had not been able to sell it. West then expressed a particular wish to see it again. The artist looked confused, and, when further pressed, reluctantly owned that, being in a state of great poverty and exceedingly disappointed by his ill-success, he had, in a fit of despair, hurled the plaster against the wall and shivered it to atoms. West, though much astonished at this indication of a phrenzied and impetuous spirit, said little at the time, but endeavoured to revive his hopes and to soothe the irritation of his feelings. Then taking his leave, he reflected much upon the incident, and on what he could do to assist and befriend this able but singular youth. The place of travelling artist to the Academy—worth then about 100*l.* per annum, I think he said—soon after became vacant, and West secured it for his young friend, thinking it was the very thing to suit him. Fearful, however, after what he had witnessed, of communicating too suddenly this pleasing intelligence, he asked him to dinner, and in the course of the evening gradually informed him of it. He was delighted, and broke forth into expressions of gratitude and hope. The sequel was such as deeply to affect West's feelings. The young artist was found a corpse in his bed the very morning after this announcement of his good fortune; and West's impression was that the sudden transition from a state of extreme destitution to one of comparative independence, acting upon his highly excited and ardent temperament, had produced this fatal catastrophe.

In the spring of this year an event occurred which spread a general gloom and consternation over the public mind, and which was keenly felt by Mr. Wilberforce on private as well as public grounds. I allude to the assassination of Mr. Perceval by Bellingham, in the lobby of the House of Commons, by a pistol wound, in the afternoon of Monday, May 11. A few days after the occurrence of this tragical event I paid a visit at Kensington Gore, and was present at a meeting between him and some of his most intimate parliamentary friends, in which it was touched upon with much feeling, and the political embarrassments with which it appeared pregnant were the subject of conversation. Mr. Stephen, Mr. Babington, and Mr. Charles Grant, among others, were present. It was observed, before the discussion closed, that this was one of those occasions in which Christians in general, but especially Christian statesmen, should offer to God their earnest prayers for a blessing upon the counsels of those who direct the affairs of the country. The remark was made with equal seriousness and propriety, and I could not but think at the moment on how much higher ground men who felt and acted thus stood, than the hackneyed politician or the blind follower of a party.

Of Mr. Perceval's personal character, and of the purity of his patriotism, Mr. Wilberforce spoke to me on this melancholy occasion in terms of high eulogy. He did not estimate his talents as a speaker as of the highest order, but said his courage as a debater was indomitable; and added, that Grattan used to compare him to a seaman who would venture out in all weathers. 'Perceval,' said Wilberforce, on another occasion, 'was a fine creature. His generosity was unbounded. I don't mean merely in giving away money, but in a much higher and larger sense. Then, there was in him such kindness of heart, so much feeling, and all based upon true religious principle. I believe he was a man of prayer. Pitt paid him a high tribute, in a conversation he had with Lord Harrowby just before he went to fight the duel with Tierney, by mentioning him as a man to whom the country might look as Minister in case he should fall.' What he said to me of his Christian and general character so closely corresponds with some passages in his own private diary, that I extract them as follows:—'Perceval had the sweetest of all possible tempers, and was one of the most conscientious men I ever knew: the most instinctively obedient to the dictates of conscience, the least disposed to give pain to others, the most charitable and truly kind and generous creature I ever knew. He offered me at once a thousand pounds for paying Pitt's debts, though not originally brought forward by Pitt, and going out of office with a great family.' ... 'Oh, wonderful power of Christianity! Never can it have been seen, since our Saviour prayed for His murderers, in a more lovely form than in the conduct and emotions it has produced in several on the occasion of poor dear Perceval's death. Stephen, who had at first been so much overcome by the stroke, had been this morning, I found, praying for the wretched murderer; and, thinking that his being known to be a friend of Perceval's might affect him, he went and devoted himself to trying to bring him to repentance.' ... 'Poor Mrs. Perceval, after the first, grew very moderate and resigned, and with all her children knelt down by the body and prayed for them, and for the murderer's forgiveness.'*

In the course of the following summer, Mrs. Harford and I were in Ireland, and spent some time in the county of Wicklow at Bellevue, the beautiful seat of Peter Latouche, Esq. Nowhere had Mr. Wilberforce warmer admirers than at Bellevue, and my intimacy with him heightened the cordiality of my reception there.

* Life, vol. iv. 26-27.

Bellevue was the centre of a circle of friends of much literary and religious interest, and it was natural that in writing to Mr. W. I should touch upon their leading characteristics. I pause here to portray a few particulars respecting them. The Rev. James Dunn, vicar of the adjoining parish of Delgany, was a cherished guest at Bellevue, and was one of the brightest ornaments of its society; but the brightest of all was the late Alexander Knox, formerly private secretary to Lord Castlereagh, and since well known by his theological writings. His intellectual and conversational powers were of the highest order. Here Mr. Grattan was also an occasional visitor, when he and Mr. Knox acted on each other's genius, and charmed everyone present by their brilliant effusions of wit and eloquence. There was a constant succession of interesting guests at Bellevue, and everyone was placed at his ease by the cordial benignity of the venerable host, Mr. Peter Latouche, and by the winning kindness of his lady, whose Christian benevolence and warmth of heart secured her universal affection and esteem. Her devotion to the interests of the Great Orphan Asylum, in Dublin, was such as to occupy much of her time and thoughts : not, however, to the exclusion of most assiduous and personal attention to the educational wants and the charitable claims of the neighbourhood. It was her delight to superintend her girls' school, and to go forth amongst the sick and poor, ministering to their religious as well as temporal necessities in the spirit of a true disciple of Christ. This most estimable pair had no children, but a niece, Miss Kate Boyle, afterwards Mrs. Hornby, of Winwick, was cherished by them as a daughter, and was every way worthy of their affection.

The theological opinions of Mr. Wilberforce differed in various important particulars from those of Mr. Knox, but he highly appreciated his spiritual and elevated views, and on the subject of Roman Catholic emancipation there was a growing accordance between them. Of Mr. Dunn he used to say: 'What an interesting creature is Dunn! he is formed of the finest biscuit.' His was, indeed, a winged and devout soul. He seemed to live on the confines of the invisible world, and there was an attraction of love, humility, and kindliness about him which won on every heart. His pulpit talents were of a high order, and their effect was increased by the noble expression of his countenance and the sweetness of his voice. His disinterested superiority, as a clergyman, on many trying occasions, to all secular considerations, was well known to his intimate friends, and secured him their reverence. Another frequent and valued visitor at Bellevue was the Rev. Robert Daly, now Bishop of Cashel, a truly zealous clergyman, whose sturdy and unflinching attachment to Protestant principles, as conspicuous then as it has ever since been, occasionally produced slight collisions between him and Mr. Knox, carried on, however, in the most friendly spirit. The Rev. John Jebb, late Bishop of Limerick, who is well known by his writings to have been a disciple as well as a friend of Mr. Knox, was also a much valued guest. I was charmed with Mr. Knox's conversation, of which Hannah More used to say, there had been nothing comparable to it since the days of Edmund Burke. His parents were converts of John Wesley, who had taken a deep interest in the developement of their son's mind in early life, and whose memory Mr. Knox cherished with the utmost reverence and affection, justly regarding him as an instrument raised up by God for promoting a wonderful revival of true religion. I one day said to him : ' Give me an idea of the character and manners of Wesley from your own recollections of him.' He replied : ' John Wesley was a human angel.' His conversation, he said, abounded in anecdotes, equally well told, instructive, and amusing. For instance, he was one day talking of the habit which many persons, even of superior education, contract of interlarding their conversation with one or another peculiar phrase, without being aware of it. Among such was the celebrated lawyer, Chief Justice Holt, whose perpetually recurring expression was, ' Lookie, d' ye see ? An admirer of the Chief Justice one day said to his nephew, 'Your uncle is a great man, but what a pity it is that he can't talk for any time together without bringing in "Lookie, d' ye see."' 'I'll break him of it,' said the nephew; and the mode he adopted was as follows: Holt had often found fault with him for not giving his mind to legal studies. One day the nephew surprised him not a little by saying: 'Well, uncle, I have thought much of your advice, and have been acting upon it so intently as to have versified parts of "Coke upon Lyttleton." Shall I give you a specimen?' Holt nodded assent, and he proceeded thus:—

> 'He that is tenant in fee
> Need neither quake nor quiver;
> For he hath it, 'Lookie, d' ye see?'
> To him and his heirs for ever.'

'Ah, you rogue,' said the old judge, 'I understand you.' Mr. Knox had quite a collection of letters addressed to him by Wesley, extracts from which he was fond of reading to such of his friends as could appreciate them.

During my stay in Ireland I again and again received interesting letters from Mr. Wilberforce, of a too private nature, however, for insertion here. He had told me in the preceding spring that he should probably resign his seat for Yorkshire at the approaching dissolution, as he found he was arrived at a time of life when the pressing business of that great county, in connection with the important questions which he had taken up in Parliament, too much absorbed his time and thoughts from domestic and other duties. I found on my return home that this step had been actually taken in September 1812, and that he had become in the following November M.P. for Bramber, one of the boroughs afterwards swallowed up in Schedule No. 1. Deep was the regret felt at his retirement by the electors of all classes, as evidenced by the resolutions passed at a public meeting of the county of York, on October 12, which bore testimony to his unremitting and impartial attention to the private business of the county, and to his independent and honest performance of his trust upon every public occasion.* We cannot omit the concluding sentence of a similar address to him from his native town of Hull: 'Amongst other subjects of praise, it is not the least that on retiring from the representation of this county, after a faithful service of twenty-eight years, and possessed of the influence which such a station must necessarily command, you have not during that period accepted of place, pension, or rank, and have acquired no other name than the distinguished title of the "Friend of Man."'

* See Life, vol. iv. p. 66.

CHAPTER II.

1813.

Renewal of the Charter of the East India Company; involving the Question, What Degree of Support to the Church of England in that Country, and to Missionaries, should be conceded by the Legislature?—Animated Debates on the Subject, in which Mr. W. took a prominent part.

IN the spring of 1813 the renewal of the charter of the East India Company came under the consideration of Parliament, and it involved the important question what degree of legislative support should be given to the branch of our national establishment in India, as also what facilities, or whether any, should be conceded to the zeal of Christian missionaries going to that country. Mr. Wilberforce will be found, on occasion of this important question, to have exercised the same energy of purpose, and the same enlightened and untiring zeal in asserting the rights of the Hindoos, as our fellow-subjects, to have the inestimable blessings of Christianity offered to them, as he did in the case of the slave trade to rescue the suffering negroes from the oppression of slavery and its concomitant moral evils. I therefore feel that a selection of a few passages from his speeches on this interesting and most important subject will not be out of place here, more especially as his efforts in this noble cause are less known in the present day, than his slave-trade exertions. I often conversed with him on the subject while he was so engaged, and caught a few scintillations of the bright flame which glowed within him as he poured forth his soul on these and similar occasions.

As far back as the year 1793 Mr. Wilberforce had moved and carried in Parliament the following resolution :—' That it is the peculiar and bounden duty of the British Legislature to promote, by all just and prudent means, the interest and happiness of the inhabitants of the British dominions in India, and that for these ends such measures ought to be adopted as may gradually tend to their advancement in useful knowledge, and to their religious and moral improvement.'

Twenty years had elapsed since this resolution had been solemnly recorded, and nothing had yet been done to render it effective. His Majesty's Government were now, however, prepared to propose a measure in reference to it, so moderate in its spirit that it seemed scarcely possible to provoke opposition. An ecclesiastical establishment, consisting of a bishop and three archdeacons, who should preside over the Indian branch of the Church of England, formed one part of the proposition, and another, though it took from the Court of Directors the despotic power of absolutely excluding the entry of Christian missionaries in India, yet subjected their proceedings while there to any restrictions which might appear prudential or expedient to the local Government. It soon, however, became apparent that these wise and beneficial measures would not pass through the Legislature without difficulty and opposition. A prejudiced clamour was raised, charging the authors of them with the insane intention of drilling the Hindoos into Christianity by parliamentary enactment; and many even of those who did not join in this clamour anticipated destruction to our Indian empire from allowing the religion of Christ to be preached in a country subject to the foul and bloody superstitions of Brahma and Vishnoo.

Public opinion throughout England anticipated the decision of the Legislature upon this interesting and important question by numberless petitions to both Houses of Parliament, which not only asserted the unprincipled injustice of systematically debarring the hundred and sixty millions of our Indian fellow-subjects from the blessings and benefits of Christian knowledge, but the positive duty of placing within their reach the means of its acquirement. It was natural that a public man, such as Mr. Wilberforce, should act upon these principles with equal zeal and firmness. He was convinced, after much enquiry, that the apprehended dangers of such a course were groundless, provided that no forcible nor undue interference were attempted with the superstitions of the natives; and experience has fully sustained the propriety of this conclusion. When, therefore, the time for action in Parliament came on, he placed himself in the van of those who opposed the mercenary policy and prejudices of a section of the East India Company, and maintained the opposite view of the question with equal argument and eloquence. In reference to this subject I received from him, before the discussion came on in Parliament, the following letter, which will illustrate the zeal and energy of his benevolence on this as on all similar occasions :—

'London: March 31, 1813.

'My dear Sir,—Though extremely pressed for time, I must send one hasty line in reply to your questions. The door has not been quite shut to missionary exertion, and I own I think it will be better to touch on the past with a tender hand, and rather to petition that for the future the system may be improved. No missionaries (but now and then, when very seldom the Bartlett's Buildings Society has had to send any) have ever been suffered to go to India in the Company's ships—the only ships, you know, which have sailed from these kingdoms for India; they have, therefore, stolen out as they could, circuitously and unlicensed. Last year a very worthy man, immediately on his arrival after a long circuit, was ordered to depart; and, on his stating that he had no money, was told he might go before the mast. The poor fellow—chiefly from grief—sickened and died. The expense of these circuitous passages is very great; but it is due to the Government abroad to say that several of the missionaries, when there, have been kindly treated. It is the Court of Directors at home that is hostile, more than the Government in India. As to petitions, I should prefer some *such* words as those of the resolution which was voted in the House in 1793, and which is contained in Dr. Buchanan's last publication.

'I remain, my dear Sir, your sincere Friend,

'W. W.'

In another letter, nearly of the same date, he thus expresses himself:—

'I am so extremely pressed for time that I must make one letter serve for Mrs. Hannah More and you. Let me beg you, after perusing the enclosed, to transmit it to her. I know your intimacy. I cannot express how deeply I feel this subject. Its magnitude far exceeds any other I remember, except, perhaps, the slave trade, and the latter equals it only in its ultimate bearings. Let me hint to you the importance of obtaining the support of as many as possible of the friends of humanity who may not agree with us in religious sentiments. All surely will join who do not wish to see such a vast body of our fellow-subjects, or rather tenants, sunk in the grossest moral, social, and domestic barbarism, not only without an effort to raise them in the scale of being, but with persevering endeavours to keep them down, and this when all the nation is earnest to make them subservient to our commercial gain. Alas! alas! Do, my dear Sir, exert yourself to the utmost.

'I am ever your sincere and affectionate Friend,
'W. W.'

On June 22, 1813, Lord Castlereagh pressed the adoption of the resolution already quoted, and it was strenuously supported by Mr. Wilberforce, in a speech of powerful argument and impressive eloquence. On that day no speaker attempted a reply, and the resolution was carried; but on a resumption of the question on the 1st of July, Mr. Marsh, in an elaborate and eloquent speech, opposed the adoption of the resolution as fraught with imminent danger to British interests in India. Nothing could be more uncandid than the style of his attack. He represented the object of the resolution to be wild and visionary, and fraught, if adopted, with peril to the safety of our Indian empire; pronounced a studied panegyric on the purity of the Hindoo code of morality, and the virtuous lives of the natives of India, though it arose in evidence before the House, on the high authority of such men as Lord Teignmouth, Mr. Grant, and others, who stated from personal observation and experience, acquired during a long residence in the country, that nothing could possibly be more debasing than the Hindoo superstition—nothing lower or more gross than the standard of Hindoo morals. Mr. Marsh proved how hard he was driven for effective argument by indulging in a great deal of random fire against cant, hypocrisy, and Calvinism—terms by which he hoped to enlist prejudice and ridicule on his side against the pressure of his opponent. Mr. Wilberforce was most prompt and able in his reply; and, while he demolished the arguments of Mr. Marsh, triumphantly established the truth and rectitude of his own positions. The peroration of his speech on this occasion was as follows:—

'And now, Sir, if I have proved to you, as I trust I have irrefragably proved, that the state of our East Indian empire is such as to render it highly desirable to introduce among the natives the blessings of Christian light and moral improvement; that the idea of its being impracticable to do this is contrary alike to reason and to experience; that the attempt, if conducted prudently and cautiously, may be made with perfect safety to our political interests; nay, more, that it is the very course by which those interests may be most effectually promoted and secured—does it not follow from these premises, as an irresistible conclusion, that we are clearly bound —nay, imperiously and urgently compelled—by the strongest obligations of duty to support the proposition to which I now call upon you to assent? Its only fault, if any, is that it falls so far short of what the nature of the case requires. Is it that we should immediately devise, and without delay proceed to execute, the great and good and necessary work of improving the religion and morals of our East Indian fellow-subjects? No; but only that we should not substantially and in effect prevent others from engaging in it. Nay, not even that; but that we should not prevent Government having it in their power, with all due discretion, to give licenses to proper persons to go to India and continue there, with a view of rendering to the natives this greatest of all services. Why, Sir, the common principles of toleration would give us much more than this. It is toleration only that we ask. We utterly disclaim all ideas of proceeding by methods of compulsion or authority. But surely I need not have vindicated myself from any such imputation—the very cause which I plead would have been sufficient to protect me from it. Compulsion and Christianity! Why, the very terms are at variance with each other—the ideas are incompatible. In the language of Inspiration itself, Christianity has been called "the law of liberty." Her service in the excellent formularies of our Church has been truly denominated "perfect freedom;" and *they*, let me add, will most advance her cause who contend for it in her own spirit and character.'

To the alleged purity of the sacred books of the Hindoos, and to the assertion that as a nation they were distinguished for virtue and morality, Mr. Wilberforce thus forcibly replied :—

'It has often been truly remarked—particularly, I think, by the historian of America—that the moral character of a people may commonly be known from the nature and attributes of the objects of their worship. On this principle we might have anticipated the moral condition of the Hindoos by ascertaining the character of their deities. If it was truly affirmed of the old Pagan Mythology that scarcely a crime could be committed the perpetrator of which might not plead in his justification the precedent of one of the national gods, far more truly may it be said that in the adventures of the countless rabble of Hindoo deities you may find every possible variety of every practicable crime. Here, also, more truly than of old, every vice has its patron as well as its example. Their divinities are absolute monsters of lust, injustice, wickedness, and cruelty; in short,

D

their religious system is one grand abomination. Not but that I know you may sometimes find in the sacred books of the Hindoos acknowledgments of the unity of the great Creator of all things; but just as, from a passage of the same sort in Cicero, it would be contrary alike to reason and experience to argue that the common Pagan Mythology was not the religion of the bulk of mankind in the ancient world, so it is far more absurd and groundless to contend that more or fewer of the 33,000,000 of Hindoo gods, with their several attributes and adventures, do not constitute the theology of the bulk of the natives of India. Both their civil and religious systems are radically and essentially the opposites of our own. Our religion is sublime, pure, and beneficent; theirs is mean, licentious, and cruel. Of our civil principles and condition, the common right of all ranks and classes to be governed, protected, and punished by equal laws is the fundamental principle. Equality, in short, is the vital essence and the very glory of our English laws. Of theirs the essential and universal pervading character is inequality—despotism in the higher classes, degradation and oppression in the lower.'

In a passage preceding that quoted above he notices, in the fifth report of the Committee on East Indian Affairs, the singular proposition of Mr. Dowdeswell for the correction of the enormities of Indian morals. This gentleman proposed that the institutions of Mohamedanism, aided by 'some remains of the old system of Hindoo discipline still existing,' should be moulded into a course of instruction for the community.

Upon this proposition Mr. Wilberforce remarks as follows :—

'We are led irresistibly by this passage to a conclusion, which I confess has been suggested to me by various other circumstances, that in the minds of too many of our opponents Christianity and India are inconsistent and incompatible ideas. We cannot but be reminded of the expression of a former ornament of this House, that "the Europeans were commonly unbaptised in their passage to India." I will not presume to adopt so strong a position; but Mr. Burke himself could not have desired a stronger confirmation of his assertions than some with which we have been supplied in the course of these discussions—more especially with this, wherein we find that a gentleman of intelligence and respectability, long resident in India, bewailing such a dissolution of the moral principle as rendered it difficult for the frame of society to hold together, and looking round solicitously for some remedy for the evil, never so much as thinks of resorting to Christianity, but proposes to resort to a revival of Hindooism and Mohamedanism as the only expedient to which it is possible to have recourse.

'Agreeing with him in my sense of the virulence of the disease, I differ entirely with respect to the remedy; for, blessed be God! we have a remedy fully adequate and specially appropriate to the purpose. That remedy, Sir, is Christianity, which I justly call the appropriate remedy; for Christianity then assumes her true character, no less than she performs her natural and proper offices, when she takes under her protection those poor degraded beings on whom philosophy looks down with disdain, or perhaps with contemptuous condescension. On the very first promulgation of Christianity it was declared by its great Author as " glad tidings to the poor ; " and, ever faithful to her character, Christianity still delights to instruct the ignorant, to succour the needy, to comfort the sorrowful, to visit the forsaken. I confess to you, Sir, that but for my being conscious that we possessed the means of palliating, at least, the moral diseases which I have been describing, if not of effecting their perfect cure, I should not have had the heart to persevere in dragging you through the long and painful succession of humiliating statements to which you have been lately listening. For, believe me, though I trust that to many in this House I scarcely need vindicate myself against such a charge, that it is not to exult over the melancholy degradation of these unhappy people, or to indulge in the proud triumph of our own superiority, that I have dwelt so long on this painful subject; yet I must be allowed to say that my great object is to impress you with a just sense of the malignity of their disease, that you may concur with me in the application of a suitable remedy—for I again and again declare to you a remedy there doubtless is. It would be blasphemy to believe that the Almighty Being to whom both we and our Indian fellow-subjects owe our existence has doomed them to continue for ever incurably in that wretched state of moral depravity and degradation in which they have hitherto remained. No, Sir! Providence has provided sufficient means for rescuing them from the depths in which they are now sunk; and I now call upon you to open the way for their application: for to us, Sir, I confidently hope, is committed the honourable office of removing the barrier which now excludes the access of Christian light, with its long train of attendant blessings, into that benighted land, and thus of ultimately cheering their desolate hearts with the beams of heavenly truth and love and consolation. And therefore, Sir, I indignantly repel the charge which has been unjustly brought against me, that I am bringing an indictment against the whole native population of India. And what have they done to provoke my enmity? Sir, I have lived long enough to learn the important lesson that flatterers are not friends—nay, that they are the deadliest enemies. Let not our opponents, therefore, lay to their souls this flattering unction—that they are acting a friendly part towards the Hindoos. No, Sir, they—not I—are the real enemies of the natives of India, who, with the language of hollow adulation and "mouth honour" on their tongues, are in reality recommending the course which is to keep those miserable beings bowed down under the heavy yoke which now oppresses them.

'Not, Sir, that I would pretend to conceal from the House that the hope which, above all others, chiefly gladdens my heart is that of being instrumental in bringing them into the paths by which they may be led to everlasting felicity; but still, were all considerations of a future state out of the question, I hesitate not to affirm that a regard for their temporal well-being would alone furnish abundant motives for our endeavouring to diffuse amongst them the blessings of Christian light and moral instruction.'

In the following passages he depictures in a striking manner the singular, the marvellous features of an Indian empire, and the security which would be imparted to it by the conversion of its inhabitants to the Christian faith :—

'On the most superficial view, what a sight does that empire exhibit to us! A little island obtaining and keeping possession of immense regions, and of a population of sixty millions that inhabit them, at the distance of half the globe from it! Of inhabitants differing from us as widely as human differences can go—differences exterior and interior—differences physical, moral, social, and domestic—in points of religion, morals, institutions, language, manners, customs, climate, colour: in short, in almost every possible particular that human experience can suggest or human imagination devise! Such, Sir, is the partnership which we have formed: such, rather, the body with which we are incorporated—nay, almost assimilated and identified. Our Oriental empire, indeed, is now a vast edifice; but the lofty and spacious fabric rests on the surface of the earth without foundations. The trunk of the tree is of prodigious dimensions, and there is an exterior of gigantic strength. It has spread its branches widely around it, and there is an increasing abundance of foliage and of fruit; but the mighty mass rests on the ground merely by its superincumbent weight, instead of having shot its roots into the soil and incorporated itself with the parent earth beneath it. Who does not know that the first great storm might possibly lay such a giant prostrate?' After enlarging at some length in the same strain, he goes on to say: 'If, Sir, we would render ourselves really secure against all such attacks, let us endeavour to strike our roots into the soil, by the gradual introduction and establishment of our own principles and opinions—of our own laws, institutions, and manners—above all, as the source of every other improvement, of our religion, and, consequently, of our morals. Why, Sir, if it were only that we should thereby render the subjects of our Asiatic empire a distinct and peculiar people—that we should create a sort of moral and political basis in the vast expanse of the Asiatic regions, and amidst the unnumbered myriads of its population—we should render our East Indian dominions by this change more secure, merely from the natural desire which men have to preserve their own institutions, solely because they are their own, from invaders who would destroy them. But, far more than this, are we so little aware of the vast superiority even of European laws and institutions—and far more of British laws and institutions—over those of Asia, as not to be prepared to predict with confidence that the Indian community which should have exchanged its dark and bloody superstitions for the genial influence of Christian light and truth would have experienced such an increase of civil order and security, of social pleasures and domestic comforts, as to be desirous of preserving the blessings it should have acquired; and can we doubt that it would be bound by the ties of gratitude to those who had been the honoured instruments of communicating them?'

Considering the frequent difficulties to which our Indian empire is exposed, this speech, so full of wise counsels and important facts, deserves to be thus specially referred to.

CHAPTER III.

1813 TO 1815.

Mr. W.'s Delight in the Beauties of Nature—His Hints as to the Best Modes of preparing for Parliamentary Speaking—Madame de Staël seeks his Acquaintance—Retreat of the French Army from Moscow, and great Political Events which followed, ending in the Peace of 1814—Disappointment of Mr. W. at the Want of Securities in that Treaty for the Abolition of the Slave Trade—His striking Appeal in Allusion to it—Lord Castlereagh's Reply—Presence of Crowned Heads and Illustrious Foreigners in London at this Epoch—Mr. W. waits on the Emperor Alexander by his express Command—Public Meeting at Freemasons' Tavern with respect to the Defects of the Treaty of Peace as to the Slave Trade—Splendid Illuminations in London—Mr. W. visits them—Duke of Wellington a cordial Abolitionist—Death of Mr. Thornton and of various Friends.

IN the early part of 1813 I experienced a renewal of Mr. Wilberforce's hospitality and kindness; and in the following autumn we had the happiness of receiving him and Mrs. Wilberforce, with their two daughters, as our guests for several days at Frenchay, where we then resided. It proved a truly interesting visit, and in the course of it I made him acquainted with several excellent and benevolent individuals whom he had pleasure in meeting. Among these was Richard Reynolds, a justly venerated philanthropist of the Society of Friends, whose striking countenance, fine patriarchal figure, and simple cordial manners, much pleased him; and who, on his side, was delighted with the opportunity of seeing and knowing William Wilberforce.

During this visit I witnessed for the first time his enthusiastic admiration of the beauties of nature. We were traversing a deep wooded glen, which forms the approach to Blaise Castle, whither I was conducting him to call upon my father, and nothing could exceed the pleasure he expressed in surveying its scenery. I may be excused for adding that my parents, who had long revered his name and sympathised in his exertions for the abolition of the slave trade, gave him the most cordial welcome, and greatly prized his visit. In the course of the many visits which in after-years he paid me at Blaise Castle his first impressions of its scenery never failed to be revived; and many of its walks and drives have acquired in my esteem a peculiar charm from their intimate association with the image, conversations, and the partialities of my revered friend.

The conversation happening on one of the days of his stay with us to turn upon public speaking, Mr. Wilberforce made some forcible remarks upon the influence which a man of ability and judgment may acquire by cultivating this talent. Thence he proceeded to dilate upon speaking in Parliament and the best mode of preparation for it. Were he giving counsel, he said, to a young member, he would particularly caution him against courting applause at the outset of his career, by ambitiously aiming to make what is called a *fine speech*. Should the attempt prove successful, such an undue estimate might probably be formed of the speaker's abilities as would render his subsequent and less studied efforts failures. Or should he unfortunately break down—a case by no means uncommon under such circumstances—vexation and disappointment might possibly seal his lips for ever. There was no better preparation, he added, for the style of speaking most adapted to the House, than a diligent attendance on committees, and a careful attention to the details of business and evidence which come before them. A great fund of useful practical knowledge on various important topics might thus be acquired, which would qualify a man, whenever the reports of such committees became the subject of debate, to supply the House with what it especially valued—accurate and useful information. The discussions carried on in committees frequently resembled in every particular, excepting the excitement of a great popular assembly, the debates of the House itself. By frequently taking a part in these, a man of any ability for speaking would soon acquire the habit of reasoning and expressing himself correctly and with parliamentary tact. He had known many gentlemen who, though labouring at first under much embarrassment and difficulty, had thus successfully made their way, and risen at length into consequence and consideration. To aim at a logical arrangement of the ideas, and to cultivate the habit of elegant and correct writing, were also essential to success. These were points to which Mr. Pitt used to direct the attention of young speakers, whom he also recommended to commit to memory a few striking thoughts with reference to any debate in which they proposed to take a part, in order to have something ready to retreat upon, in case of difficulty or nervous embarrassment. He had himself, he added, found it a useful practice in his younger days to engage a friend to read aloud to him suitable passages out of any distinguished author, and then to repeat them as nearly as possible in the same words. Of Mr. Pitt he said that he came into Parliament so accomplished an orator, that in the arrangement of his matter, the force of his reasoning, and in all the graces of finished elocution, his first speeches were almost equal to his last. Mr. Windham's speeches, he said, were known to have been prepared with assiduous care, and, though interspersed with anecdotes which seemed spontaneous, to have been written down before delivery; and Sheridan's were so diligently elaborated, that he had often been known, before the occurrence of a great debate, to shut himself in his room, day after day, where he was heard declaiming for hours together. Of the rhetorical treatises of the ancients he gave the preference to Tully, 'De Oratore,' which well deserved to be carefully studied, as an admirable epitome of the science of speaking.

At this time Mr. Wilberforce had been deeply interested by a MS. letter which had been shown him, addressed by the Rev. R. C. Whalley, rector of Chelwood, to a friend of his, on the subject of Communion with God; so much so, indeed, that he had obtained permission to have it copied out for his own private use. He was already well acquainted with the name of Mr. Whalley, and felt the highest esteem for his Christian character, which was so remarkable that we think our readers will not be indisposed to listen to a few particulars concerning it, which, together with some passages from the above-mentioned letter, will be found in the last chapter of this volume.

In the early spring of 1814, Madame de Staël was in London, with her son and daughter, where she mingled much in the highest society. Having expressed a particular wish to become acquainted with

Mr. Wilberforce, the Duke of Gloucester, in order to gratify her, made up a select dinner-party, to which he was specially invited. She afterwards made a great point of his dining with her, which he did, and, in addition to her son and daughter, the company included Lord and Lady Lansdowne, Lord Harrowby, Sir James Mackintosh, &c. One of the guests afterwards assured me that, without any effort on Mr. Wilberforce's part, the conversation was quickly in his hand, and that such was the brilliancy of his thoughts and remarks, that Madame de Staël herself, after thus meeting him, said, 'I have always heard of Mr. Wilberforce as amongst the most benevolent of men : I shall now ever think of him as one of the wittiest and most agreeable.'

He told me that he had afterwards sent her his book on 'Practical Christianity,' for which she had almost asked; and her remark on it to a mutual friend was, 'C'est l'aurore de l'immortalité.'

The destruction which overwhelmed the French army under Napoleon in the autumn and winter of 1812, amidst the snows of Russia, was followed by the co-operation of the Great Powers of Europe to break his iron yoke. The capture of Paris by the armies of the Allied Powers, and the victories of the Duke of Wellington in the north of Spain and at Toulouse, diffused throughout this country in the spring of 1814 unbounded joy and thankfulness. The spell of

Napoleon's imaginary invincibility had been signally broken—his despotism over the Continent was at an end: and Great Britain, which, through a long struggle of twenty years, and for the most part single-handed, had maintained her own independence, and frustrated the devices of his boundless ambition, was hailed as the great champion of the liberties of Europe. In feelings of cordial joy at these signal successes, and in thankfulness to the Almighty Disposer of events, no heart beat higher or warmer than that of Mr. Wilberforce; but it will be seen by the following letter that he hoped to extract from these events matter of other triumphs than those of national exultation over a baffled foe :—

'Kensington Gore: April 20, 1814.

'My dear Sir,—Your letter did not reach me till the afternoon of yesterday; and to-day I was called from home early, and was too much occupied after my return to take up my pen. Indeed, I am just now engrossed in time and thoughts by considering how to render the great events which have lately taken place, and the negotiations which will soon follow, subservient to the grand purpose of a general convention for the abolition of the slave trade by all the great European nations. This, however, you will consider as confided only to yourself for the present. A premature disclosure of our purpose might call forth opposition which would not otherwise be made.'

He then touches, as follows, on a case of charity which I had brought before him:—

'I am sorry you felt any difficulty or delicacy in applying to me respecting the school at ———. My own conduct towards you should have prevented those sensations; but they arose from your friendly consideration for me, and on that ground they claim and receive my gratitude.

'I wish I knew what sum it is right for me to contribute. Do, as a friend, decide for me. I am aware that (independently of other considerations) something is due to character in these cases—the motive, for instance, of letting your light so shine before men, &c. (Matt. v. 16), though I own I have formerly felt at a loss in adjusting the claims of this principle with those of the opposite one, of not letting your right hand know what your left hand doeth. Is not the right medium first to purify our practical principle as much as possible, and then to give publicly all that is fairly to be expected from a man of your known or understood fortune, circumstances, &c., and what you may do in any case more than this to do privately? I shall be quite satisfied with whatever sum you may determine. While I continue in public life my saving a shilling is out of the question.

'I entirely sympathise with you in the late events. Never was the hand of the Almighty displayed more strikingly. Your reference is exact. "Howbeit he thinketh not so."

'Believe me, &c. &c.

'W. W.'

The hopes of Mr. Wilberforce, and it may be said of the country, were severely disappointed when the terms of the treaty of peace between England and France, then governed by her legitimate monarch, Louis XVIII., were made public. France, instead of entering into a pledge, on receiving back her ancient colonies in the West Indies, and her factories on the coast of Africa, for an immediate abolition of the slave trade, had obtained license, and England was made a party to the compact, to carry it on for a term of five years; and no positive or definite terms whatever had been entered into with Spain or Portugal on the subject. When, therefore, Lord Castlereagh appeared in the House of Commons on June 6, bearing under his arm the treaty of peace and amity with France, and the House, as was most natural, burst forth into universal and enthusiastic cheers at the sight of a document which attested the glorious termination of the fierce and protracted struggle which Great Britain had maintained for her national independence—nay, for her national existence *—the only voice which remained silent amidst this burst of joy was that of Mr. Wilberforce. No heart beat more highly than his with patriotic emotions on an occasion so deeply interesting; but this feeling was mastered by another which forbade its utterance. After Lord Castlereagh, amidst these expressions of general exultation, had moved that the treaty do lie on the table, Mr. Wilberforce rose, and assured his noble friend that if he had not been able to concur in the salutation with which he had been welcomed on his return to the House of Commons, it was not from any want of personal cordiality, but because, seeing the noble lord coming up to the House with the French treaty in his hand, and calling to mind the arrangements made in it respecting the slave trade, he could not but conceive that he beheld in the noble lord's hand the death-warrant of a multitude of innocent victims—men, women, and children —whom he had fondly indulged the hope of his having himself rescued from destruction. To witness, therefore, the revival of this dreadful evil, when to so great a degree he had conceived it was extinct, could not but fill his mind with the deepest grief and disappointment; and as for the stipulation that the French themselves would join in abolishing the trade in five years, at the period of life to which he

* My description of this scene is quoted in the Life, vol. iv. p. 187.

had arrived, with the experience he had gained, with the historical and diplomatic knowledge he had collected, he could not be at all sanguine that this stipulation would take effect; for if *now*, when the French had no capital engaged in the slave trade— not a ship, not a merchant, not a manufacturer—they yet cleaved so closely to this abhorred system, how much more must we fear that they would value and cling to it when they would have so strong and manifest an interest in adhering to it. After touching further on this topic with great force, and expressing his deep regret that out of an occasion so auspicious and favourable nothing more effectual should have arisen, he urged on the noble lord the importance of at least applying some principles of limitation to the extent and manner of carrying on the trade. He feelingly alluded, in conclusion, to his own disappointment, and said that his noble friend must allow for the acuteness of the feelings under which he spoke, on reflecting that when, at length, after a laborious contention of so many years, he had seemed to himself in some degree in possession of the great object of his life, when the cup was at his very lips, it was rudely dashed from them for a term of years, if not for ever.

Lord Castlereagh, in reply to this eloquent appeal, candidly owned that he felt very great regret at having been obliged, as he conceived, to consent to the only terms on which France would have been willing to bind herself to act at a definite period in concert with this country for the attainment of the great cause to which his honourable friend had so nobly devoted all his energies. He thought, however, that his honourable friend took far too desponding a view of the consequences of the stipulation which he had reprobated, for he was himself persuaded it was the sincere intention of the French Government faithfully to redeem the pledge into which it had entered.

The sentiments expressed by Mr. Wilberforce on this occasion were warmly re-echoed by numerous addresses to Parliament, which poured in from every part of the kingdom: and the general impression of the public mind was, that a great moral principle, which might have been successfully maintained, had been yielded to diplomatic complaisance and foreign influences. That such was the case will hereafter be confirmed by the authority of one who, from his commanding situation, could not be deceived on such a question.*

During part of the time that these discussions were in progress I was staying with Mr. Wilberforce at Kensington Gore. London was just then in a state of the greatest excitement and enthusiasm from the presence of the Emperor of Russia, the

* The Emperor Alexander.

King of Prussia, and a throng of illustrious foreigners, both statesmen and warriors, who had assembled there after the first capture of Paris to exchange their congratulations with the Prince-Regent of England and the British nation at the auspicious termination of the mighty contest which had so long shaken Europe to its centre.

I was present at a dinner party at Mr. Wilberforce's house at this time when a note was delivered to him. Shortly after reading it he handed it to me, whispering to me at the same time to make no allusion to its contents. It was a note from one of the officers of State attendant on the Emperor Alexander, expressing his Imperial Majesty's wish that Mr. Wilberforce would wait upon him the next day, at an hour which was named. On returning it to him I expressed in a low voice the pleasure it gave me, to which he replied that he hoped to render the interview subservient to the furtherance of the abolition cause. No hint was given by him to the company in general of the contents of the note throughout the evening—a forbearance perfectly in unison with his simplicity of character.

The hour named by the Emperor being an early one, Mr. Wilberforce appeared at breakfast the next morning in his court dress, and I shall never forget the impressive manner with which he told me how earnestly he had been imploring the blessing of God upon the approaching interview. Alexander, he afterwards informed me, received him with a kind and cordial frankness, which at once made him feel at his ease, for as he was about to kneel and kiss his hand, the Emperor caught him by his own hand and shook Mr. Wilberforce's in a most friendly manner. After a little general conversation, Mr. Wilberforce led him to the subject of the slave trade, and expressed his regret that the recent congress should have separated without any stipulations for its immediate abolition. The Emperor expressed himself as being personally most favourably inclined to such a measure, and hinted that more might have been effected by British influence had it been efficiently exercised by Lord Castlereagh. After Mr. Wilberforce had earnestly commended the cause to the future support of his Imperial Majesty, the conversation turned upon Russia, and the Emperor touched on the imperfect state of civilisation in a great part of his own dominions, and his anxiety to promote the improvement of his people. They parted with mutual esteem.

At a subsequent period the King of Prussia sent Mr. Wilberforce a present of beautiful Berlin porcelain, as a mark of his grateful recollection of the eloquent manner in which he had advocated the claims of the north of Germany—wasted and devastated by the closing struggles of the war—upon British benevolence. Blucher also gave a remarkable proof of the consideration with which he regarded the bloodless triumphs of the British statesman, by charging the messenger whom he despatched to England with his own report of the battle of Waterloo with a letter of a similar tenor to Mr. Wilberforce.

About the time of which I am now speaking, viz. the spring of 1814, I one day accompanied him to a public meeting at the Freemasons' Tavern, Great Queen Street, the object of which was to address Parliament and the Prince-Regent to amend, as far as might be possible by the interference of the British Legislature, the defective enactments of the recent treaty of peace with respect to the abolition of the slave trade. The great room was full to overflowing when we arrived, and the proceedings had already commenced. All the leading members of the Opposition, including Lords Grey, Holland, and Lansdowne, and Messrs. Brougham, Tierney, &c. were present. There was also a large attendance of those who were mainly prompted by their benevolent feelings to step forward at this juncture. Mr. Wilberforce was recognised as soon as he entered the room, and a lane was quickly formed for him to reach the platform. As we advanced, the meeting began to cheer him; but for a few moments he was quite unconscious that he himself was the object of applause, for walking with his head declined upon his breast he saw no one. As he leant on my arm he whispered to me with perfect simplicity, 'Have you caught what is going on?' 'They seem to me,' I replied, 'to be all cheering you.' We reached the platform just as I had uttered some such words, when in an instant the fact was signally confirmed, for the moment he was placed in a conspicuous position the whole room rang for some minutes with repeated thunders of applause. In this manner I have seen his presence hailed on many occasions of a public nature, and I cannot conceive it possible for any human being, either at the moment or subsequently, to have been less acted upon by any particle of vanity under such demonstrations of esteem, or rather to have been in a greater degree superior to them. Anything like elation was checked by a deep and intimate principle of Christian humility which pervaded his inmost soul. I was much struck on this occasion, as I often was subsequently in the House of Lords, by the elegant and Ciceronian style of Lord Grey's speaking; though, in point of lively and stirring effect upon the feelings of the auditory, what followed from Mr. Wilberforce was far more efficient and impressive. The meeting, before it broke up, passed a vote of thanks to him for his persevering exertions, in which they justly designated him the father of their great and noble cause.

The Emperor of Russia, during his stay in London, came to the House of Commons one day, accompanied by his sister, the Duchess of Oldenburgh, to be present at a debate. He stopped Mr. Wilberforce, whom he met in one of the passages of the house, with a manner full of kindness, and said he must have the pleasure of introducing him to his sister. The Duchess addressed him in the same friendly way, and added, that having been already introduced to him, she must now claim his acquaintance.

There were splendid illuminations and other demonstrations of public joy for several successive nights at this time in London in honour of the peace. The whole city was one blaze of effulgent light, and the ornaments and devices in coloured lamps were superb. On one of the evenings I accompanied Mr. Wilberforce and Dean Milner from Kensington Gore to the Houses of Parliament to view the brilliant scene. As we returned we stopped near Carlton House, the exterior screen of which, as well as the beautiful portico, was hung with coloured lamps, and looked like an enchanted palace. Its claim to this title, on a former night, was heightened as I approached it by the *coup d'œil* presented by the entrance-gates flying open, and a succession of the royal carriages, in all the splendour of their State liveries, rolling out one after the other, and bearing the Prince-Regent and his imperial, royal, and illustrious guests to a sight of the illuminations. Suitable escorts attended them. Great injunctions had been laid upon me to watch over Mr. Wilberforce throughout our evening expedition, and, knowing how unequal he was to stand the squeezing of a mob, I felt not a little uneasy at his expressing a particular wish, at Charing Cross, to quit the carriage and take a nearer survey. The Dean, who was as corpulent in his person as his friend was slender and aërial, wisely resolved to remain behind. Carriages in all directions were darting along—the mob, though in high glee and good humour, was in a state of the greatest pressure—and after we had once quitted the carriage it proved very difficult to regain it. Mr. Wilberforce was much amused, and appeared little aware of the difficulty I had in guiding and protecting him, for we several times narrowly escaped being run over, and in one place got jostled amongst a set of pickpockets, who were very active in plying their trade under such favourable circumstances. The Dean, on hearing of our adventures, reposed with double complacency upon his own wisdom in not quitting the carriage.

While the treaty of peace with France was pending, Mr. Wilberforce had addressed a letter to Prince Talleyrand, in which, after touching in a forcible manner upon the countless evils and the impolicy of the slave trade, he expressed an earnest hope that France, following the example of Great Britain, would adopt the policy of immediate abolition. The reply of the wily old politician was highly complimentary to Mr. Wilberforce, but, as respected the main object, was in substance as follows: England, with all her superior lights and philanthropy, did not abolish the slave trade till after deliberations prolonged through fifteen years. It is a little too much to expect of France that she should take the leap without any deliberation. In showing me this letter, Mr. Wilberforce smiled at the ingenuity of the excuse for delay, though he deeply deplored its consequences.

Throughout the ensuing summer and autumn he lost no favourable opportunity of pressing on the Duke of Wellington, then our ambassador at Paris, the views of the abolitionists, and of acting through him on the French Government. The following letters will throw light on the proceedings thus instituted:—

'Sandgate: Oct. 12, 1814.

'My dear Sir,—Your name has for several days had a place on my memorandum-paper in the list of persons to be written to. I wish to mention to you a capital French publication in favour of the abolition of the slave trade, by M. Sismondi, whose two works—the one on the Literature of the South of Europe, the other on the Italian Republics—have been spoken of in high terms in the "Quarterly Review." You will see, I hope, a considerable portion of it in a number of the "Ambigu," which I desired a friend to send you. Peltier, the author of that periodical work—which comes out, I believe, once a fortnight—has lately taken a very decided line on our side; and he has so much intelligence and so much wit, that I am desirous of promoting the sale of his work as much as I can. It has suffered of late from the return to France of a great number of emigrants who used to take it in; and the line he has taken respecting St. Domingo and the slave trade has alienated many of his former friends. You will have read in the newspapers the account of the "Red Book" of Hayti, of which island I believe the real state would, if told to the greater part of the West India proprietors, appear not a whit more credible than Swift's account of Lilliput or Laputa.

'You enquire concerning our present prospects. You will hear with pleasure that the Duke of Wellington, our ambassador, is a cordial abolitionist. But though we are assured that the King of France personally is well affected to our cause, and also that Prince Talleyrand and several others of his Ministers are favourable, yet the tide runs in general so strong against us that the Ministry profess to be unable to act up to their own sense both of duty and interest. You will be glad to hear that the Prince-Regent wrote a letter with his own hand to the King of France, assuring him that he could not give our Prince a more acceptable proof of his regard than by consenting to the abolition. The answer I have seen, promising abolition, only not too suddenly, and in the meantime a limitation of it. But, notwithstanding all this, I am very uneasy: for you must have seen stated in the Dutch papers a letter from the French Custom-house, dated August 29 last, authorising the recommencement of the slave trade under the old regulations: and we know that an armament sailed a month ago to take possession of the settlements surrendered by us on the African coast. Therefore, notwithstanding the repeated and strong assurances given to Lord Castlereagh by Talleyrand, and to the Duke of Wellington since, that the slave trade should be prohibited throughout all the district in which we should prove to them that, *in fact,* the slave trade had been discontinued, I repeat it, I am very uneasy. Yet I am not despondent. The very circumstance of there being so many eminent literary men favourable to our cause appears to me to be an indication from Providence that we may hope the best; though Chateaubriand, and Humboldt, and Madame de Staël, and Sismondi may not be aware that they are accomplishing the purposes of Heaven. "Howbeit thou meanest not so, neither doth thy heart think so." But I must say farewell, after thus gratifying your curiosity a little. I am myself intending to recover, or rather acquire, if at my time of life I can acquire, some power of speaking French. Could I have expressed myself in it with ease and elegance I should long ago, I doubt not, have been in Paris.—I am ever, with cordial esteem and regard,

'Yours very sincerely,
'WILLIAM WILBERFORCE.'

'Chelsea: Nov. 23, 1814.

'My dear Sir,—If I had not been much engaged I should have anticipated your question by informing you that a letter from the Duke of Wellington brought me a very few days ago the welcome intelligence that the French Government had actually issued an order prohibiting the slave trade by French subjects anywhere to the north of Cape Formosa. This is to the eastward of the Bight of Benin, and comprehends above 1,500 miles of coast; the more valuable because, from the shape, slaves might be brought from the interior either to the southern or the western coast. Let us praise God for this great blessing, and hope it may be an earnest of more complete success. Believe me, it always gives me pleasure to hear from you; and therefore favour

me with a letter when you have any tidings concerning yourselves or common friends to communicate, or when any spontaneous effusions from your own mind press for a conveyance. I assure you they will be welcome. Did I send you a copy of my letter to Talleyrand? and, if not, would you have it in French or English? for I have been advised to print an English edition. I assure you the prospect you paint of a continental excursion is a very gratifying one to me; but it must not be next year, for family reasons.—Believe me ever, my dear friend,
'Yours very affectionately,
'WILLIAM WILBERFORCE.'

Early in the winter of 1815, I received a letter from Mr. Wilberforce, alluding to the decease of my honoured father, of which the following is an extract:—

'February 14, 1815.

'I did not till a short time ago hear that your long course of solicitude had terminated in the death of your beloved parent.

'Even by those who feel concerning the events of this chequered life as real Christians, such an incident as the death of a parent, or even of a near and dear friend, will be felt severely; and, indeed, it ought to be so felt, for here, as in so many other instances, it is the glorious privilege of Christianity, and the evidence of its superior excellence, that it does not, like the systems of human fabrication, strive to extinguish our natural feelings, from a consciousness that it is only by lessening them that it can *deal with them*—if I may so express myself— and enable us to bear the misfortune as we ought; but it so softens, and sweetens, and increases the sensibility of our hearts and tempers as to make us love our friends better, and feel more keenly the loss for the whole of this life of our former delightful intercourse with them; and yet it, at the same time, so spiritualises and elevates our minds as to cheer us amidst all our sorrows, and enabling us on these as on other occasions to walk by faith and live by the Spirit, it raises us to the level of our ascended friends, till we hear almost their first song of exultation, and would not even wish to interrupt it, while we rather indulge the humble hope of one day joining in the chorus.

'Yet the loss of so excellent a man as Bowdler, at what seemed to us so premature a period, when we might have hoped that for so many succeeding years the world would be instructed by his wisdom and charmed by his eloquence, and, above all, edified and improved by his example, must be deeply felt by the survivors. And even in the case of Mr. Henry Thornton (I at least may naturally feel this), who was of the same age, much it might be hoped still remained for him to do for the benefit of his fellow-creatures and the glory of God.* And Buchanan, too; but—I am silent.' . . .

* Mr. Henry Thornton, Mr. Bowdler, and Dr. Buchanan, were all three removed to a better world within six weeks of each other.

CHAPTER IV.

1815 TO 1818.

Pleasant Excursions and Visits in Company with Mr. Wilberforce—Author's Residence in Rome—His Acquaintance with Cardinal Consalvi, the Papal Secretary of State—Mr. W. requests the Author to place before the Cardinal important Documents relative to the Slave Trade—The Cardinal arranges for him in consequence a Private Interview with the Pope, Pius VII.—Correspondence with Mr. W.

IN the summer of 1815, Mr. Wilberforce, accompanied by one of his sons and a young friend, spent several days with us. I went to Oxford to meet him. As we roamed from one college to another in that venerable seat of learning, or lingered in the public walks and gardens, his imagination was powerfully excited by the charm of the surrounding objects; and he seemed almost to wish at the moment to begin life over again, in order to taste the delights of a youth sedulously employed in the acquisition of knowledge amidst these shades of studious retirement. His feelings were in close unison with those of Gray in his beautiful tribute to the Gardens of the Sister University:—

F 2

> Ye brown o'er-arching groves,
> That Contemplation loves,
> Where willowy Camus lingers with delight:
> Oft at the blush of dawn
> I trod your level lawn;
> Oft wooed the gleam of Cynthia silver-bright
> In cloisters dim, far from the haunts of Folly,
> With Freedom by my side, and soft-eyed Melancholy.

He often expressed a very deep feeling of regret that the period of his own university life had been too much in the character of a blank. Happily, however, for himself and for his country, he became duly sensible of the value of time while he was yet young, and repaired in a great degree the want of studious habits at college, by assiduously pursuing an enlarged course of reading after he had laid aside his gown.

During his stay with us, we visited the romantic scenery of the Wye, with which, and especially with Tintern Abbey and its adjacent objects, he was delighted. We next spent some days at Walton Castle, with my father-in-law, Mr. Hart Davis, for whom and for the various members of his family he ever felt and manifested a most cordial regard. Hence we proceeded to the Deanery of Wells, on a visit to Dr. Ryder, Bishop of Gloucester, afterwards Bishop of Lichfield and Coventry, a friend whom he highly prized and loved—a prelate after his own heart —who united to the zeal of an apostle the most amiable and endearing qualities, and the polished manners of the best society. The north of Devon and its beautiful coast was our next object. We went by way of Dunster and Minehead; and 'The Valley of Rocks Inn,' at Linton, became our head quarters, whence we explored the romantic beauties of the vicinity.

There are few parts of England which comprehend within the same compass so much striking scenery. The Bristol Channel here begins to expand into open sea. The coast is fringed by bold and towering cliffs, wild in their general aspect, but softened in many parts by hanging woods and intervening dingles of exquisite beauty. Here and there the range of these cliffs is broken by extensive glens opening on the sea, which, when explored, unfold an enchanting combination of woods, rocks, torrents, and small rivers, whose crystal waters flowing over abrupt beds, and amidst stones studded with exquisite mosses and lichens, often break into cascades, and enliven with their murmurs the deep solitudes of these fine recesses. On the crest of one of the finest of these glens, and looking down upon the sea, Linton is situated. We devoted one day to an excursion to Water's-meet, so called from the concurrence of two of the romantic streams above described. This charming spot is now made too easy, as it was then abruptly difficult of access. We all rode thither on ponies, and dined, seated on some huge stones in the bed of one of the streams, beneath the pendant shade of over-arching trees. Mr. Wilberforce enjoyed himself in the spirit of a young man of twenty. He was brimfull of happiness and poetry, of which he read and repeated to us many beautiful passages; and we closed this delightful excursion by spending a few days with our mutual and much-valued friends, Sir Thomas and Lady Acland, at Killerton Park, near Exeter. Often and often he has said to me, '*I love Acland;*' and his affection, I may truly add, met with a warm response from a heart made up like his own of kind and generous qualities. Neither Mrs. Harford nor I can ever forget the pleasure which we enjoyed in Mr. Wilberforce's society during this excursion. His mental energy never flagged. We were often amused at the capacities of his pockets, which carried a greater number of books than would seem, if enumerated, credible; and his local memory was such that, on drawing out any author, he seemed instantaneously to light on the passage which he wanted. In addition to the stores of his pockets, a large green bag full of books filled a corner of his carriage, and when we stopped at our inn in the evening it was his delight to have this bag into the parlour, and to spread part of its stores over the table. He kindled at the very sight of books.

We parted at Killerton under the impression of a long separation, for I was on the eve of visiting the Continent, with the intention of spending a considerable time in Italy. During a residence in Rome, which extended through parts of two years, circumstances occurred which made me well acquainted with Cardinal Consalvi, the Papal Secretary of State, and I became, in consequence, the medium of some interesting communications between him and Mr. Wilberforce, with reference to the abolition of the slave trade by Spain and Portugal. Cardinal Consalvi, after discharging with great diplomatic skill the functions of Papal ambassador at the Congress at Vienna, presided over the administration of civil affairs in the Pontifical States with almost unlimited authority. His official career was signalised by the correction of many abuses, and by salutary reforms in the administration of justice; and he would have carried improvement much farther had not his plans been thwarted by the bigoted opposition of a strong party in the Conclave, who regarded him as a liberal and an innovator. His character and deportment formed a strong contrast to those of most of his brother cardinals. The personal pomp and stateliness affected by the order were his aversion, though on occasions which called for it he manifested a just sense of what was due to the dignity of his station. To simplicity of manners, and to an utter absence of personal pretension, he united unaffected politeness, and a kindness and suavity of demeanour which were peculiarly winning.

His countenance was truly striking. It equally beamed with the expression of intelligence, penetration, and benignity. His dark hazel eyes, beneath impending brows, shone with a peculiar lustre, and his figure was tall and imposing.

It so happened that about this time strenuous efforts were making on the part of England to induce the Spanish and Portuguese Governments to follow her liberal and humane policy in the abolition of the slave trade; but as they were not prepared for going thus far, it struck Mr. Wilberforce that it would be well to endeavour to engage, if possible, the Papal Government to exert its influence with them for the furtherance of this important object. He had heard from me of the easy access which I had to the cardinal, and it will be seen, in the following letters, that he was anxious that I should be the medium of placing before him various important facts connected with the subject of the slave trade:—

'The only obstinate continuers of the slave trade are now the Spaniards and the Portuguese, the nations most under the influence of the Romish Church; and instances of Spanish slave ships, and of Portuguese, also, have come to my knowledge attended with circumstances almost too dreadful to be described. In one case, a vessel of 160 to 180 tons contained, when she sailed from Africa, above

500 slaves, besides the sailors, &c.; and how they were heaped together you may judge when I tell you that probably there would not be room for even 300 of them to lie on their backs. When they were taken, several were dying daily from suffocation, and that inexpressible (by any known word) complication of miseries resulting from cramming together into the hold of a ship at sea a vast number of human beings unused to that element, in a tropical climate, the men necessarily in chains, and shut down under hatches during the night. In this dreadful situation the flux is often generated, and then, from the impossibility of the people's getting to their tubs, from the comrade who is fettered not consenting to make a simultaneous effort to secure that convenience—from the quarrels and often the bitings of the legs of those who do try to crawl over and through the jammed bodies of their fellow-victims—such a mass of filth and horror is witnessed as is insufferable, even to the coarse but honest senses and feelings of our rough sailors. In another vessel, owing to the slaves being concealed under a false deck, the whole mass of them were on their hams, and the whole floor one large tub: this, in the stormy sea off the Cape of Good Hope, was the horrid condition of a slave-ship going from Mozambique to Rio. There, also, the daily deaths attested the fatal as well as cruel nature of the sufferings of the poor

wretches. And nothing has been said of wives and children probably left behind—of being forced from their own country to be taken they know not where. Surely, surely, if the Pope, or if Consalvi were made aware of the real nature of these abominations, and of their unparalleled extent, they would exert themselves to put a stop to them. And such has been the inclination of the scale in Spain already, that a very small weight put into it would probably make it preponderate—the Council of the Indies having (at least a majority of it) recommended to the Spanish Government the entire and immediate abolition of the slave trade. How much would it be for the credit of the Roman See to take the same share in putting an end to this gigantic evil, which one of the Popes (I think Gregory the Great) had in terminating the private wars and the state of slavery in Europe in the Middle Ages. But more than this, in proportion as the Papal State should show itself active in opposing the slave trade, it would entitle itself in the estimation of all good men to the grateful as well as the humane protection of the navy of this country from the ravages of the Barbary pirates. Let me beg of you, if you properly can, to convey these ideas to some of the leading men in the Cabinet of His Holiness. But I have been trespassing on your time, and have been prodigal of my own, you will say, since it is to be feared I am too late with my suggestions, and I blame myself exceedingly for not having communicated them to you before; indeed, I have often made efforts to get His Holiness interested in our cause, but I was not aware that you could have had such access to him. Lord Castlereagh himself has never, I think, been sufficiently impressed with a sense of the importance of calling in the aid of the Roman Church. It is out of the ordinary diplomatic course.

'I will detain you no longer than while I beg my kind remembrances to your lady, and assure you that I am, with cordial esteem and regard, &c. &c.

'W. WILBERFORCE.'

The letter containing these revolting details reached me in the summer of 1816. I had spent part of it in Switzerland; but on my return to Rome in the ensuing winter, I lost no time in placing it before the Cardinal in an Italian translation, and appealing to him in the manner suggested to me by Mr. Wilberforce. He listened with the greatest kindness and courtesy, and assured me that I might depend upon his furthering to the utmost of his power the benevolent object of the distinguished individual on whose part I applied to him. 'We must take care,' he added, 'to remonstrate with Spain and Portugal in such a way as will not unnecessarily excite angry feelings towards ourselves. You know they are, of all the other European States, the most devoted to the Holy See : while, therefore, I am truly sincere in the promise of earnest interference, and hopeful of the issue, I must well consider the most politic method of exerting our influence.' As I had found that neither Consalvi, nor any of the leading members of the Conclave, were adequately acquainted with the history of the slave trade, or with the efforts of England for its abolition, I had intermediately written to Mr. Wilberforce, requesting him to send me any pamphlets or printed evidence giving useful information on these points. The following was his reply :—

'Kensington Gore: Nov. 15, 1816.

'My dear Sir,—I cannot easily express with what regret I reflect that your friendly and most interesting letter of September 24 has not yet been answered. Yet I trust I need not assure you that this delay is not to be ascribed to negligence or procrastination. The sickness of a friend, which recently called me away 200 miles and more from my family, and detained me in close attendance, and the death of my dear and only sister, Mrs. Stephen, which soon followed, together with the difficulty I have experienced in securing for my parcel a safe conveyance to Rome—all these circumstances will satisfy you that my tardiness is in no degree owing to my not esteeming the business itself as of the very utmost importance. I have, indeed, been quite at a loss how to convey the books to you. Lord Bath, however (in Lord Castlereagh's absence his substitute), has kindly promised to convey it with the Government despatches. I send you one magnificently bound copy of my Slave Trade Letter, either for Cardinal Consalvi, or for His Holiness himself, as you may judge best: I should suppose the former. I also send some copies of my letter to Talleyrand; but what may prove still more valuable, I send four copies of a pamphlet, drawn up by Lord Castlereagh's order and under his superintendence when at Vienna. This composition will probably derive an authenticity, and be regarded as entitled to a weight, from its origin, which do not belong to any of the others. And now, my dear Sir, let me encourage your hopes, and animate your exertions, by informing you that even in Spain, that *deadest* —if I may use such a word—and most torpid of all European countries, the translation into Spanish of a tract on the Abolition of the Slave Trade has produced, as I am assured on very high authority, some effect on the Court of Madrid. I reproach myself for not having been more provident and assiduous in thus diffusing information: but it is an unspeakable comfort to bear in mind that we have to do with a God whose ways (more especially in pardoning our sins, negligences, and ignorances for Christ's sake) are not as our ways, nor His thoughts as our thoughts. I need say no more on this head. The publications will tell you all, except that you will naturally insist on the disgrace which will attach to the name of Christian, and still more of Roman Catholic, if, when the Protestant Church of Great Britain and Prussia, and the Greek Church of Russia, have manifested an earnest zeal for the extinction of this system of wickedness and cruelty, the two countries which are notoriously the most under the influence of the Holy See should alone obstinately cleave to it, and even assist the natives of the other European kingdoms in counteracting the liberal and humane intentions of their respective Governments. Surely your touching this string must produce some considerable vibration. Now to your last letter. You excited strange longings for a preternatural power of locomotion, by the very date of that letter, Lucerne, Chamouny, &c., and by anticipations of Greece under the convoy of Frederic North. You could nowhere find so good a cicerone. He is one of the sweetest-tempered men I ever knew. This is, indeed, hereditary with him. But, alas! I fear continued familiar intercourse with persons of various modes of faith and worship has a sad tendency to produce that spirit which leads a man to regard them as so many roads tending to the same ultimate point of destination. Your other friends keep you so well informed concerning the state of England that I need say nothing on that head. As respects our commercial, agricultural, and manufacturing interests, I could say nothing which it would not give you pain to hear; but blessed be God, in the concerns which are of all others most important, we are going on so well as to afford, I trust, a pledge that if the Almighty chastens us, He has not given us over unto death.* The number of truly religious and zealous men among our clergy especially is continually increasing.

'Farewell! I beg my kind remembrance to Mrs. H. How enchanted she must have been in the paradise of the scenery of Lucerne.—Believe me, with every good wish,

'Very sincerely yours,

'W. W.'

In the course of various interviews with the Cardinal after the receipt of the above letter, I sought to touch the chord referred to by my excellent friend, and not wholly without success. The books and pamphlets which he had forwarded were carefully distributed by me amongst the most influential of the cardinals. I was honoured, through Cardinal Consalvi, with a private interview with His Holiness (to

* At this time there was a great depression in our commercial and manufacturing interests.

whom I had previously been presented), for the purpose of requesting him to accept a copy of Lord Castlereagh's pamphlet referred to above, splendidly bound for the occasion. My interview with the Pope took place at the palace of the Vatican. I paced for some time the Papal antechamber among numerous officials and others waiting there, with my book under my arm, and I was amused at hearing afterwards of their surmises who the savant was that was about to present a book to His Holiness. On being ushered in I found the Pope quite alone. His manner was kind and gracious, and I at once stated that I felt highly honoured in being permitted to approach him, in order to request him to accept, on the part of some illustrious men among my countrymen, who were zealous promoters of the abolition of the slave trade, a work on that subject, which I trusted would interest him. I then unfolded the print of the interior of a slave-ship, and called his attention to its revolting details—at which he expressed astonishment and pity. I pressed upon him the words of Mr. Pitt, whose name, I concluded, must be familiar to him, 'That the slave-trade was the greatest practical evil that had ever afflicted humanity.' I then expressed the hope that his potent influence would be used with the Governments of Spain and Portugal to engage them to enter into the benevolent views of Great Britain for the abolition of so horrible a traffic.

He interrupted me by saying, 'They cannot fail to adopt a similar policy.'

Shortly after, Consalvi told me that everything he had promised on the part of the Pope had been done - that a letter from His Holiness had been addressed to the Court of Madrid, strongly urging its co-operation—and that a nuncio, who was just setting out for the Brazils, carried with him very pressing instructions to the same effect.

It was impossible to look at the venerable Pope without feeling sympathy for his past misfortunes, and for the firmness he had displayed in opposition to the efforts of the first Napoleon to render him the tool of his ambition. Pius VII. was a strict monk, and his habits were most self-denying. His features were striking, and their general expression bespoke calm firmness of purpose, deep mental reflection and profound humility. Though far advanced in years, his hair was still of raven-black. He usually spoke to those presented to him in Italian, but allowed them to reply in French. I felt it a high privilege to have been admitted to his presence in furtherance of so interesting and important an object; and it will be found by the following letter from Mr. Wilberforce that he was gratified with the result of my interview:—

'London: May 7, 1817.

'My dear Friend,—I am quite ashamed of my

dilatoriness in not having yet replied to your highly interesting letter from Rome, giving me an account of your labours with His Holiness. Be assured, however, that it has not arisen from undervaluing the importance of your communication. Why, you are the very prince of negotiators. Lord Castlereagh himself could not have opened his business with more adroitness. But really, by writing in this strain, I may appear less seriously convinced than I truly am, not only that you have managed the affair extremely well, but what you will read with still more pleasure, that I trust what you have done will be of no small importance. Very providentially it happens that in Brazil, whither I presume the nuncio to the Court of Portugal will have repaired, the priests are in general the most enlightened and the most humane men in the community, and, as was to be expected from persons of this description (*et sapit et mecum facit*, you know, Horace says), they are in general far more friendly than the rest of the community to the abolition of the slave trade, and to the emancipation of the slaves themselves. I do hope, and the idea is really gratifying to me, that you will be able to reflect with pleasure that you have had an important share in effecting, through the goodness of Providence, the extinction of this largest limb of that *monstrum horrendum, informe, ingens*, that still remains to it. This is the season of anniversaries. The Bible Society meeting, two days ago, was the most interesting ever remembered. Farewell! Let me beg my affectionate remembrances to your lady. You are highly privileged in thus rambling. There is nothing I should so much enjoy. But we must all endeavour to serve our generation according to the will of God.—Ever your sincere Friend, 'W. W.'

Much of the above letter has been omitted, as relating to private matters. The ensuing session of Parliament, 1818, witnessed the publication of a treaty between England and Spain for the abolition of the slave trade by that Power on certain stipulated conditions; and though it would be too much to attach any great importance to the interference of the Romish Court in a case of this description, at least I was willing to hope that some little good might have been effected.

84   RECOLLECTIONS OF WILBERFORCE.

CHAPTER V.

1817 TO 1820.

Author's Return to England—Renewed Intercourse with Mr. W.
—Visit with him to Newgate, and Readings of Mrs. Fry to the
Female Prisoners—Visit from Mr. and Mrs. W. and Family—
Slaves sold from England into Ireland—Meetings with Mr. W.
in London, and in the Country—Walks with him at Bath, and
Anecdotes related by him of Lord North and others—The late
Lord Teignmouth, and Tribute to his Memory — Mr. W.'s
Bachelor Life in Palace Yard—His Estimate of Robert Hall
as a Preacher—Visit from Mr. W. in December 1820—Various
Anecdotes—Mr. W.'s Work, entitled 'A Practical View of
Christianity;' Remarks on, and various Extracts from it.

WE quitted Rome in the early summer of 1817, and returned home by way of Venice, Pola, Trieste, &c. At Vienna and Dresden, where we made some stay, it gratified me to hear the name of Wilberforce pronounced with veneration by two eminent persons—. one, the Oriental scholar, M. Hammer; the other, Count Harrach, a most excellent and amiable nobleman, who had devoted his time and thoughts to medical studies with so much success that he was often pressed to take part in consultations upon dangerous cases occurring amongst the higher classes of society; but this he rarely did, for it was his ambition to act exclusively as the physician of the poor. There was one hospital in Vienna to which he peculiarly devoted his attention, and he was the life and soul of it. 'When I am harassed by the vexations of life,' said he, 'I come to refresh my mind among these poor creatures.' The Count had a fine library, and the gem of it, in his esteem, was a volume which had belonged to John Howard, and which contained the name of that great and good man in his own handwriting on a blank leaf at the beginning. He showed it to me with all the enthusiasm of kindred benevolence. To both these gentlemen we were introduced by Sir Thomas Acland; and it was in the course of various visits paid to the leading charities of Vienna, under Count Harrach's auspices, that he touched with interest on the character of Mr. Wilberforce.

At Brussels I received the following letter from Mr. Wilberforce, in the course of which he refers to what I had written to him about Count Harrach:—

'London: July 23, 1817.

'My dear Sir,—I fully intended answering your letter the very morning after I received it, but I really have been compelled to delay taking up my pen for a very few days. You may perhaps suppose that now Parliament is prorogued I have the command of my own time—in which case I admit it would follow, as a consequence, that an absent friend, more especially in a foreign country, has a clear right to a liberal share of it. But, alas! as if by one consentaneous movement, when the doors of the House of Commons were closed, the flood-gates flew open which had kept in the accumulating mass of my multifarious businesses. Letters, papers, public meetings, private interviews, social and domestic claims without number—all flowed in upon me with full tide and with overwhelming power. So that I have only been released from one great creditor to become the prey of five hundred smaller, each of which, however, has a legitimate claim. Yet a friend abroad shall have the priority, though I must merely send you an assurance of the continual recollection of those you have left behind, keeping unexpressed the various reflections which arise in my mind from perusing your letters and ruminating on the circumstances in which you are placed. You, I am persuaded, will return home better prepared to enjoy and to estimate the superior blessings of your own country, yet with a love, not of that base and idolatrous kind which is bigotedly blind to all its imperfections, and full of low prejudices and antipathies towards other lands and their inhabitants; but, on the contrary, glad to bring home any exotic improvement of our own system, and still more longing and endeavouring to communicate to foreigners the benefits civil and, above all, religious which we enjoy. All this, and much more, has been called forth by what you tell me of the Pope, and of cardinals, and archiepiscopal invitations. Oh, my friend! I am in a situation to be deeply sensible of the value of the best blessings of our highly-favoured land. Mrs. ——'s mother was left a widow about a year and a quarter ago, having one unmarried daughter left to be the companion and the staff of her declining years—a daughter about 32 or 33— from her early youth favoured with a large communication of heavenly grace: having also naturally a sweet temper, a pleasing exterior, and a very good and cultivated understanding—everything, in short, to make her dear to all to whom she was most nearly related. Well, she and her old mother came out of Warwickshire about a month ago to visit the married daughter. She caught a cold, or rather increased a cold, on her journey. Her complaint soon appeared to be diseased lungs. Four days ago she burst a vessel. Her death is daily, I might say hourly, expected, and is only delayed, the medical man says, by the greatest composure he ever witnessed in such circumstances—for she is quite aware of her danger. She lies in a state of calm humble assurance of the mercy of her God and Saviour; and her mother, sister, and brother are all in that state which the affectionate regard for a very dear relative, when chastened by submission and cheered by Hope and Faith and Love, produces in the mind—the feelings not extinguished, but the painful part of them lessened to a wonderful degree by the firm confidence that the friend whom we love, when released from her poor frail body, will enter into the unfading glories of Heaven. I ought to apologise for these egotisms, yet what do we wish of an absent friend but that he should pour forth out of the fullness of his thoughts and feelings—in short, that he should become present with us, and write as in that case he would talk. Your account of Count Harrach is very gratifying. It is quite cheering to hear there are such men in foreign countries. It is one of the grand superiorities of our own that it affords the most numerous openings for those who wish to serve or improve their fellow-creatures. I could not feel comfortable if I were not to express—though I trust if silent, you would believe I uttered mentally—my kind wishes for your dear lady. Every blessing be the portion of you both here and for ever. I shall be hovering about London from the 4th to the 10th of August—our posterior destination is not fixed. We should be delighted to catch you for two or three days. Write when you can. Farewell!

'WILLIAM WILBERFORCE.'

After our return to England I did not see Mr.

Wilberforce till nearly the close of the year, when I paid him a visit, which was preceded by the following note:—

'Kensington Gore: Dec. 9, 1817.

'My dear Harford,—Unless your business requires your being a mile and a half eastward, I beg you will make Kensington Gore your head quarters. I am just now extremely occupied, and ought to be much more so—so that the only way by which we can see anything of each other will be by your being in the house. You shall be as much at your own command as to hours, and presence or absence, as if you were in an hotel. To-day the Dean of Carlisle and Daniel Wilson are to be here. We dine at a quarter before five, and without further notice shall be happy to see you.

'Yours ever,

'W. W.'

In the spring of 1818 we had the pleasure of accompanying Mr. and Mrs. Wilberforce and various other friends to Newgate, to hear Mrs. Fry read the Scriptures to the female prisoners. The portion selected by her was the 7th chapter of St. Luke. Her voice was remarkably clear and sweet, and she read it most impressively. Tears were in the eyes of many of the prisoners. She concluded by offering up a prayer adapted to the occasion.

In September 1819, Mr. and Mrs. Wilberforce and

some members of their family paid a pleasant visit to Blaise Castle, and in the following month I met him by appointment in Bristol, when I had a delightful *tête-à-tête* with him. He was in most cheerful spirits, full of playful wit, and glowing with piety and benevolence. Amongst other things he mentioned that in the year 1788 he was in so bad a state of health as to cause his friends serious anxiety, when Dr. Warren, a physician of eminence, said to some of his acquaintance, 'That little fellow, with his calico guts, cannot possibly survive a twelvemonth.' He was himself so impressed with the same idea, that he earnestly entreated Mr. Pitt, in case of his death, to take up the slave-trade question, which he engaged faithfully to do. 'What changes and revolutions,' said he, 'take place in nations and in cities. Time was when Ireland was more enlightened than England, and imported Britons as slaves into her territory. England was then a sort of Africa to Ireland. Giraldus Cambrensis, or William of Malmesbury, mentions a curious fact. A plague broke out in Ireland. The priests began to enquire what it was that had awakened the Divine vengeance, and at length concluded that it was to be ascribed to the slave trade. Bristol bore its part in the traffic. I remember the expression *funibus alligatos*. I mentioned these facts in the year 1788 to Mr. Brickdale, who was then one of the members for Bristol. He reported my words to his constituents, who quizzed him for broaching so absurd a story, and offered bets against its authenticity. When Parliament met again he came to me, and begged that I would state my authority; and well I remember walking with my folio into the House of Commons and pointing out to him the passage.' William of Malmesbury has been searched in vain for the above passage, but through the kindness of a friend I have been referred to the following statement in 'Giraldus Cambrensis, Hibernia expugnata,' Selden's edition, p. 770. A translation is added, which denounces the practice of selling the natives of England as slaves, but does not appear to be the exact paragraph quoted by Mr. Wilberforce.

'His itaque completis, convocato apud Ardmachiam totius Hiberniæ clero; et super advenarum in insulam adventu tractato diutius et deliberato; tandem communis omnium in hoc sententia resedit: propter peccata scilicet populi sui, eoque præcipue quod Anglos olim, tam a mercatoribus quam prædonibus et piratis, emere passim et in servitutem redigere consueverant, divinæ censuræ vindicte hoc eis jam redigantur.

'Anglorum namque populus, adhuc integro eorum regno, communi gentis vitio liberos suos venales exponere, et priusquam inopiam ullam aut inediam sustinerent filios proprios et cognatos in Hiberniam vendere consueverant. Unde et probabiliter credi potest, sicut venditores olim ita et emptores tam enormi delicto juga servitutis jam meruisse. Decretum est itaque prædicto concilio, et cum universitatis consensu publicè statutum, ut Angli ubique per insulam servitutis vinculo mancipati in pristinam revocentur libertatem.'

'At a synod of the clergy of the whole of Ireland, held at Ardmachia, it was stated that there was a discussion upon the subject of the introduction of strangers into the island; and it was the general opinion of all that they had deserved Divine censures on account of their sins, and principally because they had been accustomed to buy English people both from merchants, and from robbers and pirates, and to reduce them to slavery, for the English people were accustomed, even whilst their country was prosperous and under no pressure of want, to expose their own children and relations for sale, to be carried into Ireland as slaves.'

The historian remarks that both buyers and sellers deserved the yoke of slavery for being guilty of such an enormous crime.

'The aforesaid council therefore decreed *nem. con.* that the English throughout the island should be emancipated from slavery, and restored to their pristine liberty.'

In the spring of the year 1820 I had some very pleasant intercourse with Mr. Wilberforce in London, and again in the autumn I was much with him at Bath, and he with me at Blaise Castle. While staying with him at Bath, we often took long walks together, when I found his conversation, as ever, equally instructive and delightful. He one day mentioned the following anecdotes, illustrative of Lord North's habitual playfulness of mind and constant good humour:—

Once, when speaking in the House, he was interrupted by the barking of a dog which had crept in. He turned round and archly said, 'Mr. Speaker, I am interrupted by a new member.' The dog was driven out, but got in again, and recommenced barking, when Lord North, in his droll way, added, '*Spoke once.*'

On an occasion when Colonel Barry brought forward a motion on the British navy, Lord North said to a friend of his who was sitting next him in the House, 'We shall have a tedious speech from Barry to-night. I dare say he'll give us our naval history from the beginning—not forgetting Sir Francis Drake and the Armada. All this is nothing to me, so let me sleep on, and wake me when we come near our own times.' His friend at length roused him, when Lord North exclaimed, 'Where are we?' 'At the Battle of La Hogue, my lord.' 'Oh, my dear friend!' he replied, 'you have woke me a century too soon.' . . . Mr. Burke, in the course of some very severe animadversions which he made on Lord North for want of due economy in his management of the public purse, introduced the well-known aphorism—'Magnum vectigal est parsimonia'—but was guilty of a false quantity by saying vectĭgal. Lord North, while this philippic went on, had been half asleep, and sat heaving backwards and forwards like a great turtle; but the sound of a false quantity instantly aroused him, and, opening his eyes, he exclaimed in a very marked and distinct manner—'vectīgal.' 'I thank the noble Lord,' said Burke with happy adroitness, 'for the correction, the more particularly as it affords me the opportunity of repeating a maxim which he greatly needs to have reiterated upon him.' He then thundered out, 'Magnum vectīgal est parsimonia.'

Lady Glenbervie, Lord North's highly-gifted daughter, of whom I formerly saw much when in Rome in 1815, liked to talk about her father's perfect good humour, and dwelt with special pleasure on his having said to her in the latter part of his life, that in spite of his long course of political warfare he could venture to say he had not a single personal enemy.

Her son, Frederick Douglas, sat in Parliament for the family borough of Banbury, and amused us one day by telling what had formerly occurred to some recreant electors, who had ventured, though vainly, to oppose Lord North's nomination of the Mayor, shortly before the annual dinner, to which his lordship was in the habit of sending venison. The old steward, while carving it, sent plenty of fat to the obedient voters, but made the rebels feelingly sensible of his displeasure, by exclaiming, as he despatched their respective plates, 'Those who didn't vote for my lord's Mayor, sha'n't have none of my lord's fat.'

Mr. Wilberforce, referring to Lord North's son, Frederick (afterwards Lord Guilford), said he was one of the sweetest tempered men he had ever known. Such indeed he was, and I will just add that one day he greatly amused me, when we were talking of Parliamentary speaking, by saying, 'I once attempted to speak in Parliament, and it was not unnatural when I rose that my family name should at once fix every eye upon me. I brought out two or three sentences, when a mist seemed to rise before my eyes; I then lost my recollection, and could see nothing but the Speaker's wig, which swelled, and swelled, and swelled, till it covered the whole House.

I then sank back on my seat, and never attempted another speech, but quickly accepted the Chiltern Hundreds, assured that Parliament was not my vocation.'

In the autumn of this year, the late Lord Teign mouth was our guest, and mentioned that while Mr. Wilberforce was living in Palace Yard, he one day called upon him by appointment to convey him to a meeting to remonstrate against the abuse of oaths, as administered at the Custom House on very trifling occasions, and to bring the subject before Parliament. 'I found Mr. Wilberforce extremely unwell,' said Lord Teignmouth, 'and the day was most uncongenial—a biting north-east wind. Mrs. Wilberforce and Dr. Milner said to him, "It will be nothing short of madness for you to go out to-day." His eldest daughter, also, was so dangerously ill that they thought she could hardly survive the day. He said, " I feel I ought to go. I have been the principal cause of convening this meeting, and if I do not go there will not be time to bring it before Parliament this session, and the measure may be entirely lost." He went, spoke admirably, and was the life and soul of the meeting, and while it lasted seemed absorbed in the subject. I accompanied him home. When in the carriage, his anxieties respecting his child returned, and he became *all the father*; and as we approached his house his emotion was great.

Never was my heart more delighted than when on arriving I heard Mrs. Wilberforce calling from above, " Wilber, our child is better ! " '

I cannot mention the name of Lord Teignmouth without adding that he was held in high and cordial esteem by Mr. Wilberforce. He was the first president of that invaluable institution, the British and Foreign Bible Society, which, at its outset, had to encounter much opposition; but his excellent judgment, seconded by his winning candour, his perfect self-command, and his amiable temper, which nothing seemed to ruffle, much contributed to carry it through every difficulty. He survived Mr. Wilberforce nearly a twelvemonth, and his memory is justly embalmed in the affectionate veneration of a large circle of friends and contemporaries.

Mr. Wilberforce resided in Palace Yard for some time, with his friend, Mr. Henry Thornton, as bachelors, where they kept an almost open house for members of Parliament. About three o'clock daily their friends began to drop in on their way to the House, and partook of a light dinner, the number often amounting to seventeen or twenty. Lord Eldon was not unfrequently one of the party. 'It delighted us,' said Mr. Wilberforce, ' to see our friends in this way, especially as it gave us the opportunity of talking upon any important points of public business without any great sacrifice of time. Those who came in late put up with a mutton-chop or beef-steak. The Duke of Montrose called in one day as we were thus employed, but declined taking anything. Seeing, however, so many around him busy with the knife and fork, he said, "I cannot resist any longer," and down he sat to a mutton-chop. "Ah, Duke!" said I, "if your French cook could see you now, he would be quite affronted."'

After his marriage he continued to live for some years in Palace Yard in the same style of friendly hospitality.

In the November of this year I went with Mr. Wilberforce, while spending a day with him at Bath, to hear the celebrated Robert Hall, of Leicester, preach. The subject of his sermon was Prayer. Mr. Wilberforce afterwards spoke to me in the highest terms of his powers as an orator, and made some remarks which are more fully expressed in an extract from a letter addressed by him about the same time to Mr. Hart Davis:—

'Bath: Nov. 24, 1820.

'Harford and Mr. Battersby came over to hear Hall, of Leicester, and we were all delighted with him. He is such a man as I could wish some of our political friends to hear who may disrelish the same truths substantially when delivered with less depth of thought and force of diction. I really cannot place him below any of our late great men, and those, with the exception of Canning, perhaps, I rate far higher than our greatest statesmen of this day, I mean for intellectual and oratorical powers.'

After the sermon, which was most able and eloquent, we had some pleasant talk with Mr. Hall in the vestry of the chapel. It partly turned upon the genius and acquirements of the late Sir James Mackintosh, whose conversational powers they both agreed were of the highest order. Hall remarked that the facility with which Sir James commanded admiration by this ready coin had perhaps too much rendered him content with present and momentary triumphs, while he possessed all the qualifications for acquiring durable fame. Some years after this conversation, meeting Sir J. Mackintosh in a friend's house, I turned the conversation upon Robert Hall, when he spoke of the depth and solidity of his intellectual powers and of the splendour of his eloquence in terms that proved how mutual was the admiration which these two eminent men entertained for each other. Sir James spoke of Hall, not only with admiration, but affection, and mentioned that they had been fellow collegians at Aberdeen early in life, when they had formed a lasting friendship. 'We used, one winter,' said he, 'to rise at four in the morning to read Greek together. Hall had taken up Plato, I Herodotus. There was a third individual in the college who rose at the same early hour and joined us—not to read Greek, but to make coffee, which he every morning prepared for us on the understanding that we should praise it as the best that we had ever tasted. Our friend, however, after this had been duly done, often found time hang rather heavy, which even the episode of bringing in the coal-scuttle for us could not altogether remove.'

In December of the same year Mr. Wilberforce and his youngest daughter passed some days with us very pleasantly. She remarked to us twice that she thought she had never seen her father enjoy himself more, or seem in better spirits than during this visit. In fact, there was a perpetual sunshine on his countenance, and his conversation was rich in anecdotes of past times. In the course of conversation he was asked what was the latest hour to which he ever remembered the House of Commons sitting? He said eight o'clock. On that occasion, he stated, various speakers had made apologies for detaining the House, it being near seven o'clock, when Mr. Pepper Arden rose and said : 'Mr. Speaker, I shall not, like many of the honourable members who have preceded me, trouble you with apologies for rising to address the House at the present late, or rather early hour, for seven is my usual hour of rising.'

Some one mentioning a young man who, while on the Continent, prided himself on speaking no language but his own, Mr. Wilberforce replied : 'A distinction which he possesses in common with every chimney-sweeper in the country.' He said of Windham that he was the pleasantest talker he had ever known. After he was appointed Secretary for Ireand, and before he went thither, he was in a party of friends of whom Johnson was one. Windham talked upon his new office, and the abuses connected with it, adding that he really knew not how he could possibly submit to them. Johnson smiled and said: 'My dear fellow, you do yourself an injustice. Depend upon it, you will make a very pretty rogue.'

There was a judge who was famous for giving useful and affecting charges. In the course of his duty he was addressing an offender with much energy and feeling, who seemed strangely insensible. The fact was the man was nearly stone deaf. At last the culprit leant forward and said to his nearest neighbour: 'What does the judge say?' 'Oh!' vociferated the other, 'he says you will be transported for fourteen years.' 'Just about what I expected,' replied the man very coolly.

On the night that Lord George Gordon's mob crowded to the House of Commons with their petition, great agitation prevailed. All the passages leading to the House were filled with their dense masses, and it was much feared that they would burst in. Lord George himself was in the House, and several times changed his seat; but wherever he went or sat he always found General Murray, a relation of the Duke of Athol, seated next him. Murray was a dark determined-looking man. After the danger was over, Lord George, turning round and still seeing the General close at his side, exclaimed : 'General Murray, what do you mean by thus haunting me all the evening?' 'I'll tell you, Lord George,' was his reply: 'I was resolved, had the mob broken into the House, to have instantly run you through.'

Mr. Wilberforce felt much the death of Christophe, King of Hayti, which occurred in December 1820. Christophe had for some time past been applying to him for counsel and assistance in various plans for the improvement and happiness of his people. He had shown great liberality in particular in advancing money for their education, having transmitted a very liberal sum to Mr. Wilberforce to pay schoolmasters, &c.

Mr. Wilberforce had taken up with great zeal this object, so closely connected with his own philanthropic efforts for the benefit of the blacks. He had shown me, at different times, parts of his correspondence with Christophe, which I think had been chiefly carried on through the medium of Count Limonade, one of his principal functionaries. Mr. Wilberforce had a high opinion of the character and principles of Christophe, and had convinced himself, by means of enquiries which he had instituted, that the reports which had been circulated injurious to his reputation were, in general, devoid of foundation.

Early in our acquaintance, Mr. Wilberforce presented me with a copy of his admirable work entitled 'A Practical View of Christianity,' &c., and on various occasions I conversed with him upon it, and on the motives which had induced him to write and publish it. Amongst these he laid great stress on the fact that—living as he necessarily did much in the world, and among men of ability and rank in Parliament, for many of whom he felt sincere regard, and whom, nevertheless, he had reason to consider as in a state of much neglect and unconcern with respect to the 'one thing needful' —he could not but feel an earnest desire frankly to express to them this conviction, and to place before them a clear statement of what he conceived was the essence and the object of true religion. Occasions suited to such a statement scarcely ever occurred, and were, in fact, rendered impossible by the courtesies of life and the forms of society. He felt, therefore, anxious to relieve his conscience on this point, and he did so by the publication of this work, which brought before its readers in a clear manner his own views on the important subject in question. Hence, also, might be inferred the secret springs of those differences which existed in his habits and conduct in many ways from the world around him.

In this work the author places in a very striking point of view the essential differences which exist between Christianity as taught in the New Testament, and Christianity as practised in the world.

He shows that the grand radical defect in the system of mere nominal Christians is, that they either overlook or disregard the peculiar doctrines of the religion which they profess: such as the corruption of human nature—the atonement of the Saviour—and the converting, sanctifying influences of the Holy Spirit.

He shows, on the other hand, upon the clearest Scriptural evidence, that a consecration of the heart to God through faith in Christ, and a life daily guided, governed, and directed by the spirit and the precepts of the Gospel, is the duty and the happiness of all sincere believers.

To live unto Him who died for them, and daily to implore the light and grace of the Holy Spirit to enable them to do so, forms, in fact, the vital principle of true Christianity. His object was to urge his readers to a practical adoption of these fundamental principles. They were not the creed of any particular sect or party, but a clear and unquestionable inference from the teaching of Jesus Christ. They were also in conformity with the doctrines of the Established Church.

How admirably are these great principles epitomised in the following passages, selected from the work itself!

'Let us turn our eyes to Christians, who have not only assumed the name, but who have possessed the substance and felt the power of Christianity; who though often foiled by their remaining corruptions, and shamed and cast down under a sense of their many imperfections, have known in their better seasons what it was to experience its firm hope, its dignified joy, its unshaken trust, its more than human consolations.'*

'But the nature of that happiness which the true Christian seeks to possess, is no other than the restoration of the image of God in his soul: and as to the manner of acquiring it, disclaiming with indignation every idea of attaining it by his own strength, he rests altogether on the operation of God's Holy Spirit, which is promised to all who cordially embrace the Gospel. He knows therefore that this holiness is not to precede his reconciliation with God, and be its cause; but to follow it, and be its effect. That, in short, it is by faith in Christ only that he is to be justified in the sight of God: to be delivered from the condition of a child of wrath and a slave of Satan; to be adopted into the family of

* Practical View, p. 96.

God; to become an heir of God, and a joint heir with Christ, entitled to all the privileges which belong to this high relation; here, to the spirit of grace, and a partial renewal after the image of his Creator; hereafter, to the more perfect possession of the Divine likeness, and an inheritance of eternal glory.*

A work founded on principles like these, and composed in a style of equal energy and eloquence—proceeding, also, from the pen of an eminent statesman, the friend of Mr. Pitt, and the zealous advocate of every noble sentiment and benevolent institution— naturally attracted general attention, and had a most wide circulation. It was published † on April 12, 1797, and within a few days was out of print. In half a year five editions (7,500 copies) had been called for. In 1807 it was eagerly read in India. In 1826 fifteen editions had issued from the press in England. In America the work was immediately reprinted, and twenty-five editions sold in the same period. It was translated into five European languages.

The strictness of its principles, and its uncompromising condemnation of a great deal that the world approved of and practised, naturally procured for its author in many quarters among the gay and

* Practical View, pp. 277, 278.
† Life, vol. ii. p. 205.

thoughtless, the title of Saint, Methodist, &c.; but I have heard him say that he almost felt too indifferent to this species of obloquy. The love of God, the love of Christ, and love for the souls of his fellow-creatures, were his inspiring motives. He therefore spoke out almost in the spirit of one of the prophets of old, and committed the event to God, who signally blessed his faithful servant's endeavours; and he was rewarded by the approving suffrages of a large portion of the bishops and clergy, and also by the esteem and veneration of the best part of society.

The following testimony is contained in a letter to him from the learned and excellent Dr. Beilby Porteus, Bishop of London:—'I am truly thankful to Providence that a work of this nature has made its appearance at this tremendous moment. I shall offer up my fervent prayers to God that it may have a powerful and extensive influence on the hearts of men, and in the first place on my own, which is already humbled, and will, I trust, in time be sufficiently awakened by it.' The Rev. John Newton (the friend of Cowper the poet) wrote to him in a similar strain. But none of these commendations more deeply touched his heart than a message which was conveyed to him by Dr. Lawrence from that truly great man, Edmund Burke, thanking him for the instruction and comfort he had derived from his book. It is stated, on the testimony of Mrs.

Crewe, who was staying with him, that his two last days on earth were chiefly occupied in its perusal. Testimonies of a similar nature, both from clergy and laity, continued to follow him throughout life. He showed me in later years a letter from Dr. Chalmers, in which that eminent divine assured him of the spiritual benefit which he had himself derived from it. The following may probably have been the passage to which I refer: 'May you be spared to spend amongst us a long old age of piety and peace. May you still have many days of rest, and of rejoicing on the borders of Heaven. And may that book which spoke so powerfully to myself, and has spoken powerfully to thousands, represent you to future generations, and be the instrument of converting many who are yet unborn.' *

'That he acted up,' is the judgment of a shrewd observer, 'to his opinions, as nearly as is consistent with the inevitable weakness of our nature, is a praise so high that it seems like exaggeration; yet, in my conscience, I believe it, and I knew him well for at least forty years.'†

How truly his life and practice exemplified the spirit of the admirable work on which we have

* Life, vol. v. p. 294. Rev. Dr. Chalmers to W. Wilberforce, Esq. Jan. 22, 1828.
† Entry on a blank page of the 'Practical View,' by J. B. S. Morritt, Esq.—Life, vol. ii. p. 205.

been commenting may be seen, amongst other proofs, in the following extract from his biography, as to the mode in which he passed his Sundays:—

'His affections were naturally lively, but it was not to this only that he owed the preservation, all through his busy life, of their early morning freshness. This was the reward of self-discipline and watchfulness—of that high value for the house of God, and the hours of secret meditation, which made his Sundays cool down his mind, and allay the rising fever of political excitement. Sunday turned all his feelings into a new channel. His letters were put aside, and all thoughts of business banished. To the closest observer of his private hours he seemed throughout the day as free from all the feelings of a politician as if he had never mixed in the busy scenes of public life.'*

In pursuance of objects like these, so noble, so elevated, so divine, it may well be said, using his own expressive words:—

'Surely an entire day should not seem long amidst these various employments. It might well be deemed a privilege thus to spend it, in this more immediate presence of our Heavenly Father, in the exercises of humble admiration and grateful homage; of the benevolent, and domestic, and social feelings, and of

* Life, vol. iv. p. 43.

all the best affections of our nature, prompted by their true motives, conversant about their proper objects, and directed to their noblest end; all sorrows mitigated, all cares suspended, all fears repressed, every angry emotion softened, every envious or revengeful or malignant passion expelled; and the bosom, thus quieted, purified, enlarged, ennobled, partaking almost of a measure of the heavenly happiness, and become for a while the seat of love, and joy, and confidence, and harmony.'*

* Practical View, p. 168.

CHAPTER VI.

1821 TO 1824.

Various Meetings with Mr. W. at Kensington Gore and elsewhere—Dinner to meet Sir Walter Scott at Clapham at Sir Robert Inglis's—Change of Residence from Kensington Gore to Marden Park—Mode of Life there—Mr. W. under Domestic Affliction at the Close of 1821—Christian Happiness in Death in the Case of Lady —— —Stay at Marden Park in 1822—Baron de Staël—Mr. W.'s Visit to Norfolk: Acquaintance and Friendship with the Gurney Family—Mr. Buxton's Speech on Slavery—Death and Character of Mr. Charles Grant—Bishop Heber's Departure for India—Mr. W.'s Thoughts as to retiring from Public Life, 1824—His temporary Residence near Uxbridge—The Author revisits Ireland and Bellevue—Stays with Mr. W. at Bath; and receives him and Mrs. W. for about a Fortnight at his own Residence.

IN February 1821, we visited Mr. Wilberforce at Kensington Gore, and afterwards met him at the houses of mutual friends. Amongst these was a most pleasant dinner-party at Sir Robert Inglis's at Battersea Rise, whither Mrs. Harford and I conveyed Mr. Wilberforce. We met there Sir Walter Scott, Lord Sidmouth, Sir Thomas Acland, Mr. Robert Grant, Mr., now Dean Milman, Captain, afterwards Sir Edward Parry, &c. &c. Mr. Wilberforce much enjoyed this meeting with Sir Walter, and the pleasure appeared to be mutual. In the course of the evening, Sir Walter described the splendid effect produced by the illuminations of Edinburgh at the Peace, in a manner so spirited and picturesque that we several of us agreed there could be no doubt who the 'Great Unknown' was. The author of Waverley at that time still wore his mask.

I saw much of my reverend friend in London in the course of that spring. He was at that time contemplating a change of residence from Kensington Gore to an abode in the country, in order to secure more leisure, and for the quiet enjoyment of his family. Marden Park, in Surrey, formerly the residence of Mr. Hatsell, was rented by him for this purpose. He thus describes it :—' It was once a fine place, and is one of the prettiest spots that I ever saw without water—the form of the ground most beautifully varied, and the wood still fine.' The house was somewhat in the style of a French château, and contained handsome and excellent accommodation. In the course of the summer I wrote and asked him whether the increased leisure he now enjoyed had been improved by him for the prosecution of an object which he had for some time entertained, and which I had often pressed him to pursue, namely, 'Recollections of his Friend, Mr. Pitt.' I extract the closing sentences of his reply, as indicative of his deep Christian humility :—

'Marden Park: Sept. 30, 1821.

' Do you remember the passage, I think in Terence, *Commemoratio quasi reprobatio est*? Such is the effect of your mention of my increased leisure and its effects. Of leisure I have enjoyed less than I expected; but partly from infirmities, more I fear from faults, I have done nothing of the kind you mention, and very little of any kind. For this and all my other manifold sins, and my sad unprofitableness (God knows I say it from the heart and with a deep consciousness of its truth), I can only fly for mercy and pity to Him, who is not only the propitiation for our sins, but the kind and gracious sympathiser with our infirmities. O my friend, what cause have we for thankfulness who know thus where to go for pardon and for peace, as well as for grace and strength! May I be enabled to spend any remainder of life that may be spared me more profitably—more according to the measure of the rich abundance of mercies and blessings for which my utmost services, as well as warmest gratitude, are due. But I must stop. Farewell, my dear friend, and believe me, with friendly remembrances to Mrs. Harford, and every kind wish,

' Ever sincerely and affectionately yours,

' WILLIAM WILBERFORCE.'

After quoting this passage addressed to myself, I cannot resist adding the following interesting picture of his life at Marden Park, which is partly in his own words and partly in those of his sons:*—'I am profiting, I trust, from the quiet life I lead at this sweet place.' Never, surely, was family religion seen in more attractive colours. 'I only wish,' said a college friend, who had been visiting two of his sons, 'that those who abuse your father's principles could come down here and see how he lives.' It was a goodly sight. The cheerful play of a most happy temper, which more than sixty years had only mellowed, gladdened all his domestic intercourse. The family meetings were enlivened by his conversation —gay, easy, and natural, yet abounding in manifold instruction—drawn from books, from life, and from reflection. Though his step was less elastic than of old, he took his part in out-of-door occupations— climbing the neighbouring Downs with the walking-parties, pacing in the shade of the tall trees, or gilding with the old man's smile the innocent cheerfulness of younger pastimes. 'The sun was very hot to-day, and the wind south; but under the beech trees on the side of the hill it was quite cool. Dined by ourselves, and walked with the boys in the evening.'

19th.—'Gave ale and cricket to the servants and

* Life, vol. v. p. 103.

all the family in honour of the coronation. Thought it safer to refuse the invitation of a neighbour, lest my plan of quiet should be rendered more difficult.'

'How little,' he said on another day, 'does that child know how much it is loved! It is the same with us and our heavenly Father: we little believe how we are loved by Him. I delight in little children: I could spend hours in watching them. How much there is in them that the Saviour loved, when He took a little child and set him in the midst! their simplicity, their confidence in you, the fund of happiness with which their beneficent Creator has endued them; that when intelligence is less developed and so affords less enjoyment, the natural spirits are an inexhaustible fund of infantine pleasure.' *

At the close of this year his feelings and those of his family were deeply affected by the death of his beloved eldest daughter, after a long and trying illness, which she bore with true patience, sustained by those consolations which Christianity alone can impart. A letter to myself, dated January 11, 1822, gives a most pleasing picture of her state of mind:—

'Kensington Gore (at Mr. Stephens'): Jan. 11, 1822.

'My dear Friend,—Both my time and my eyes—the latter especially—have had claims on them so

* Life, vol. v. p. 103.

much greater than I could satisfy, that I abstained from writing to many of those friends who I was sure would be so kind as not to misconstrue my silence. You, however, and Mrs. Harford, were prepared by my former letter to expect the issue. Blessed be God, in proportion as the trial became sharper the cordials graciously administered to us by Providence became greater; and towards the last especially, the spirituality, the resignation, the patience, the humility combined with faith in the mercies of God, through the Redeemer and Intercessor, were delightful evidences of the state of mind of her who had been "made meet to be a partaker of the inheritance of the saints in light." It was very striking to see so timid a creature looking the King of Terrors in the face without shrinking. She expressed to her mother, indeed, her surprise that she could be so composed. The day before she died, having caused all but my wife and me to withdraw, she poured forth, with a solemnity and pathos that could not be surpassed, a beautiful prayer for herself and for us who were to survive her. And on the morning of the day on which she died, she desired her favourite Nancy to ask Dr. Black, then in the adjoining room, if there was any hope of her recovery; but if not, she sweetly added, *All is well!*

'Your remarks are truly Christian and just. May we all, my dear sir, be enabled to profit from this dispensation! I was sure you and dear Mrs. H. would feel for and with us. May every blessing be your portion.

'Ever yours truly,
'W. WILBERFORCE.

'P.S. Mrs. Wilberforce was for some days very indifferent. I thank God she is now better, and the rest pretty well.'

The following letter to Mrs. Harford will show what a truly Christian mother Mrs. Wilberforce was, and how deeply she felt on this trying occasion :—

'Friday, Jan. 11, 1822.

'My dear Mrs. Harford,—You knew our dear child, and therefore know a little what we have lost. Her motto might well be that of the lady mentioned by Madame de Genlis, who, taking a violet half hid under the grass for her seal, wrote under it—"Il faut me chercher." Those few only who knew a little of this dear child, by seeing something of her character in private and domestic life, can have the least idea how sweet a flower it has pleased God to transplant to the Paradise above. You, and those who knew and appreciated her, will always be dearer to me on earth for the love they bore to her. The chastisement is indeed severe—the bereavement irreparable. But "Oh! the thought that thou art safe," as Cowper says, "that thought is bliss." And a poor sinful worm, such as I am, has only cause to bless God that He so mercifully sustained her timid spirit—enabled her to lay firm hold on the Rock of Ages—and, while she committed her own soul in eloquent and fervent prayer to her Saviour and broke out into expressions of confident hope, finished with intercessions for her parents, brothers and sisters, and dear friends whom she left behind. His grace is indeed sufficient for us, and His strength is made perfect in weakness. How can I do anything but praise Him? He has taken her to Himself, to taste those rivers of pleasure which are at His right hand, and enabled her to tell us where and to whom she was going. Blessed be His name! Do, my dear friend, pray for us—that our other five may tread in the same path, and that all of us may meet in glory. And may our friendship not be of that kind which shall only endure to this side of the grave! May we be *Christian friends*, then, in a better world, in better society—in that society which our dear child is enjoying—shall we meet to part no more, but to join in every song of praise and every angelic employment—where the Lord God and the Lamb shall be in the midst of us, and all sorrow and sighing shall flee away. Farewell, and may every blessing attend you.

'Your much obliged and affectionate

'B. A. WILBERFORCE.'

I may add, in connection with the above interesting letters, that on various occasions Mr. Wilberforce detailed to us somewhat similar facts which had occurred among his friends; and, in particular, he more than once feelingly described the case of a lady of high rank, who was arrested by fatal illness in the midst of all that can make this world brilliant and happy. She was married to a husband most deeply attached to her—was extremely handsome and greatly admired. She was the mother of a fine family, and everything seemed to smile upon her. The illness alluded to occasioned her going to Madeira for the benefit of a milder climate, where she died in 1806. Before she left England, Mr. Wilberforce had given her a copy of that excellent book, Doddridge's 'Rise and Progress of Religion,' with which she was so much interested that in her illness it was always near her. Religion, in fact, had taken possession of her soul; and from her dying bed she sent, as a keepsake, a sprig of diamonds to a beloved niece, accompanied by this message: 'Few people have been so happy in this world as I have. I have had everything to make me happy, and I have enjoyed it to the full; yet never, in my brightest days, was I half so happy as upon this bed of suffering and of death.' 'How striking,' as Mr. Wilberforce justly added, 'was this testimony from one to whom every stream of this world's enjoyments and gratifications had paid tribute!' One of her most earnest desires was to receive the Sacrament, but there was no English clergyman on the island. Before her departure, however, a vessel arrived, having one on board, and thus her last wish was gratified.

In July 1822, Mrs. Harford and I spent some delightful days with Mr. and Mrs. Wilberforce at Marden Park, on which occasion Baron de Staël passed the Sunday there, to which his sister, the late Duchess de Broglie, has particularly referred in the interesting sketch which she has published of her brother's life. We were all very pleasingly impressed by his simple and amiable manners, and he made himself very agreeable to Mr. Wilberforce by his intelligence, and by his description of the impression produced on him by this his first visit to England. Nothing amongst us, he said, had struck him so much as the high degree of liberty we enjoyed, in union with so much respect for aristocratic institutions. Such a union, he felt, imparted great coherence to our Constitution, and was to be found in no other country in Europe. After church, he and Mr. Wilberforce took a long and retired walk together, when he opened his heart to him on religious topics, and left him with a pleasing impression that the Baron was truly seeking, and he hoped would find, the 'pearl of great price.' Events have since proved that this hope was well founded, for though prematurely cut off in the flower of his days and usefulness, he has left behind him a character highly and justly esteemed for his intellectual qualities and truly Christian virtues. He assured Mr. Wilberforce that he daily read the Bible with prayer to God. He spoke with pleasure of the writings of Dr. Chalmers, and made many very acute and sensible observations on the relative position of political parties in England. At a few miles distance from Marden Park is situated the vicarage and parish church of Godstone, the scene of the pastoral labours of an old and valued friend both of Mr. Wilberforce and of myself, the Reverend Charles Hoare, Archdeacon of Winchester. His preaching and the general tenor of his ministrations were highly appreciated by Mr. Wilberforce, and the society of himself and Mrs. Hoare added much to his social enjoyment. We greatly admired the scenery of Marden Park as we wandered about in company with its interesting occupant, who himself took us to his most favourite points in the beautiful grounds which surrounded the mansion—lively was his rural enthusiasm, and delightful the buoyancy of his spirits.

During one of our visits to Marden Park, the subject of newspaper misrepresentation having been mentioned, Mr. Wilberforce said, 'Have you ever seen the " Scotsman"? It is written with uncommon talent, but is thoroughly Radical. It has often made severe attacks on me, but on such occasions a copy of the paper has been always sent me. It accuses me, amongst other things, of being the greatest jobber living. To such statements I can truly say I am wholly indifferent. They do not reach the man, for the fact is, that during the last twenty-five or thirty years, I have never asked a favour either for myself or for a relation.'

Mr. and Mrs. Wilberforce were to have visited us, but were obliged to put it off. In the month of October I received from him the following letter:—

'Elmdon House, nr. Coventry : Oct. 23, 1822.

' My dear Friend,—I have been delaying to answer your friendly letter, in the hope of finding an interval of leisure, when I should be able to do it in somewhat of a less hurried way than that in which I must otherwise have replied to it. But day passes away after day, and that hoped-for interval does not arrive. On the contrary, it seems to be farther off than before ; and therefore, without waiting any longer, I will thank you for your letter, and assure you that to hear a good account of you both must always give us sincere pleasure. You are very kind in giving me an account of some of our friends within your circle. Our Barley Wood * friend is a wonder to all of us. I

* The residence of Mrs. Hannah More.

heard from Macaulay two or three days ago (Mrs. Wilberforce being now at Barley Wood) that she sees a constant influx of visitors, and is as animated and unsubdued as ever; nor does she appear to be fatigued or exhausted by her exertions. Mrs. Wilberforce and I, with our family, partly for the health of the former, stayed for some little time at the sea-side at Cromer, a place the neighbourhood of which could not but be interesting to me, since it afforded me the opportunity of roaming among the hereditary oaks of poor Windham, and visiting his library, looking into his books, &c. &c. There is something very affecting in such a visitation. There is also in that part of the world a set of intelligent and excellent men, who have formed of late years a most valuable accession to my acquaintance, and I trust friendship. I refer not only to Buxton, but to the numerous family of the Gurneys. The young one of that name (John Joseph), who continues a plain Friend, is a most interesting character—of extraordinary piety, considerable learning, and of a delicacy and tenderness that endear him to all the wide circle of his associates. He is the more interesting from having been left a few months ago a widower, with two sweet children, to lament the loss of an excellent woman to whom he had been united a very few years. This is one of the mysterious dispensations of Divine Providence to which it becomes us to bow with humble reverence, and with a confidence in the Divine wisdom and goodness of Him who has a right not only to give but to take away. We afterwards spent a very pleasant ten days with another friend; but we were somewhat impatient to come to this place, three years having now elapsed since we had visited Mrs. Wilberforce's aged mother, now seventy-six. Dr. Chalmers spent a day or two with us, and was really the delight of us all. He expounded in the family in a most pleasing way, and his language was highly impressive, without anything of the verbosity and prolixity with which his printed sentences are somewhat charged. I dare not enter into any discussion of his plans relative to the Poor Laws, except that he appears to me—as often happens in human affairs where people are engaged in a good cause—to be prosecuting a purpose in which he will certainly produce no little public benefit, whether he succeeds or not in carrying the main object he has in view. The case might not be unnaturally likened to that mentioned, I think in the "Spectator," where the physician advised a patient to beat about a hollow ball in which certain medical infusions had been lodged, the real benefit being to be derived from the exercise. He profited from the process, but not in the way he had expected.

'You are in a happy circle of friends, but we need no other inducement to bend our steps towards your quarters than those which Blaise Castle itself affords.

It is now, however, so late in the year that I fear we must banish the hope which I assure you we had often indulged, ever since we set out upon our rambles, of looking in upon you before we should return home.

'I have not touched at all upon public affairs, which are now, in truth, but another name for private ones. I everywhere receive most distressing accounts of the state of the agricultural interest; but I dare not enter on such a field, and therefore will only express my hopes that the gracious Providence, which has so often interfered for our deliverance, will still bless us as a nation at a time when we are certainly doing more than ever before for the extension of true religion upon earth.

'I remain ever, my dear Friend,
'Very sincerely yours,
'W. WILBERFORCE.'

On January 17, 1823, while paying a most pleasant visit to Marden Park, where I had much confidential intercourse with Mr. Wilberforce, I was painfully summoned away by a family affliction, which is feelingly alluded to in the close of the following note:—

'Marden Park: Feb. 10, 1823.

'My dear Friend,—I have in vain been hunting for the letter you mentioned. Indeed, I remember carrying it in my pocket to show it to friends till its *integrity* was impaired. I will send it if I should

yet succeed. It was a younger pair of legs than
mine which jumped out of a pony cart, and produced
the kind enquiries of so many friends. I was all the
time quietly at this place. Your account of your
dear sister-in-law's closing scene is truly delightful.
How sincerely may we say, "May our last end be
like hers!" Mrs. W. has been indifferent for some
days, and I doubt if she will be able to go to town,
as was intended, to-morrow. But (D.V.) we shall
settle there soon. I hope your lady is well, and
begging you to assure your brother that I think of
him with real interest and sympathy,

'I am, ever affectionately yours,

'W. WILBERFORCE.'

Early in March 1823, Mr. Wilberforce had published a strong appeal to his countrymen, of all
grades, to give him their united support in furtherance of measures which he was anxious should be
enforced by Parliamentary sanction for improving
the condition of the slaves, in order to prepare them
for the enjoyment of liberty. So wise, and so powerful at the same time, was the spirit of his appeal on
this subject, that a West Indian proprietor told him
—'Its perusal has so affected me, that should it cost
me my whole property, I surrender it willingly, that
my poor negroes may be brought, not only to the
liberty of Europeans, but especially to the liberty of

Christians.' During the spring of this year he was far from well, and from failure of strength committed the leadership of the great cause of slavery emancipation to his friend, Mr. Fowel Buxton, who had long rendered him able and energetic assistance in the prosecution of this object. Mr. Buxton made his first speech in this character on May 15.

I often had occasion, when staying in London, to observe the singleness of heart, and the striking superiority to all merely earthly motives of conduct, manifested by Mr. Wilberforce, whether I met him in the lobby of the House of Commons, which I sometimes did by appointment, or otherwise, or within the precincts of his own dwelling. He was precisely the same man in the bustle of the one and in the quiet of the other; and I suspect that seldom has a letter been dated from that House more illustrative of that assertion than the following to myself:—

'House of Commons: April 25, 1823.

'My dear Friend,—Do not measure by the tardiness of my reply the force of the feelings excited by your last friendly and affecting note. I did not observe, till some time after, that it was written in London. I set about a letter to the Bishop of Calcutta, which is nearly finished, and I will send it to Mr. Hart Davis's to-morrow, or Saturday. The most affecting part of your letter I shall like to talk to you about. O my friend! you struck a string which vibrates in my heart in full unison. When I review all my past life, and consider ever since it has been my general desire to live to the glory of God, and in obedience to His laws, what have been my obligations, and what ought to be the amount and effects of my gratitude—what my means and opportunities of usefulness, and what the scantiness of my performances, and with what alloy their motives have been debased—alas, alas! my friend, I have no peace, no rest, but in the assurances of pardon and acceptance to the penitent believer in Christ Jesus; and I adopt the language of the publican, with the blessed declarations of mercy and grace held out to the contrite and broken-hearted. What a blessed truth it is, that it is our duty to be confident in the undeserved bounty and overflowing loving-kindness of our heavenly Father. Farewell!

'Ever affectionately yours,

'W. WILBERFORCE.'

In October of this year he lost his old and much-valued friend, Mr. Charles Grant, so long chairman of the Board of Directors of the East India Company, whose services to British India he justly deemed of inestimable value. He alludes to that event in the following letter:—

'Elmdon House: Nov. 7, 1823.

'My dear Friend,—I have given you too much

cause to scold me; but really I can boldly plead—
Not guilty. The truth is, that if I could have been
satisfied with sending you such a letterling as I must
now despatch, you would have heard from me long
ago; or if I could have been content to dictate to
you. But I felt I had much to say, and therefore
always awaited a season of leisure. It was very kind
in you to give us hints for our Welsh tour. Friendly
regard is often more truly shown by little than by
great acts of kindness. But all this time I am thinking of Mrs. H. May it please God to restore her to
perfect health. Alas! my dear friend, though we
know the shortness and brittleness of human life, we
too seldom feel it, and are quite surprised, as well as
grieved, when it is brought home to our hearts by
such an incident as I now deplore, in the sudden
death of my dear and excellent friend, Charles Grant.
I can truly say he was one of the very best men I
ever knew; and had he enjoyed in early youth the
advantages of a first-rate education he would have
been distinguished in literature as he was in
business.

'He lived so habitually above this world as to be
ready to be called at any moment into the next; and
I cannot but believe that it was in mercy he was
spared all those bodily sufferings which were likely
to be very great in the dissolution of a frame so
firmly compacted as his. He will be a great public

loss, though we are called on to praise the goodness of God in not taking him from us till he had been enabled to sow and cultivate the good seed in India, so as to ensure, humanly speaking, a large and continually augmenting harvest. But I must stop. Farewell!

'I am ever your sincere Friend,
'W. W.'

On visiting town in the spring of 1824, I found Mr. Wilberforce so unwell as to be disqualified for enjoying the society of his friends. He had been very severely indisposed. Before I quitted town, he was so much better as to receive small knots of his particular friends to take tea with him. He was in a pleasant house at Brompton. I spent a delightful evening in his society, and met there two highly-valued friends, the Bishop of Lichfield and the late Mr. Richard Ryder. We found him in animated spirits, and his heart glowing with all its accustomed kindness. Talking of the best English authors, he observed: 'In my younger days I was often in company with the first Lord Camden. He was partial to me, for I delighted in his conversation, and was sedulous in paying him those little attentions which are always grateful from young men to those more advanced in years. He was full of anecdotes of the olden time, and fond of literary talk. " For pure racy English," said he to me one day, "read South's sermons; for argument, read Chillingworth."'

As the limit of my stay in town was nearly reached when we thus met, I did not see my revered friend again till late in the ensuing autumn, but I heard from him in September. The firmness of his handwriting, and the animation of his style, verified the gratifying assurance of his improved health and strength.

Mr. Wilberforce was prevented by the slavery question from attending the Christian Knowledge Society, to hear the valedictory address of Dr. Kaye, Bishop of Bristol, to Bishop Heber, on the eve of his departure to India. I was myself present on that most interesting occasion, and, had it been possible, should gladly have gone there in his company.

I had written to him in the summer of 1824, suggesting whether the state of his health during the preceding spring, and his increasing inaptitude for the labours of Parliament, should not induce him to consider whether for his own sake, and that of his family, he would not do well to entertain the question of retirement from public life. His reply was as follows; and I the more gladly insert it on account of the high and just tribute which it contains to his and my revered friend, the late Mrs. Hannah More.

'Near Uxbridge: Sept. 11, 1824.

'My dear Friend,—Were my correspondence to

be regulated by my feelings, your last truly kind letter would have received a much earlier reply. You may perhaps have forgotten its contents, but they are fresh in my recollection, and will long continue so. You kindly gave me, but in a more Christian-like dialect, Horace's wise advice,

<blockquote>Solve senescentem maturè sanus equum, &c.</blockquote>

and, believe me, I clearly recognise the wisdom as well as the kindness of your counsel—though I scarcely dare trust myself on the topic of my wish that any remaining particle of efficiency might be employed more than my faculties have as yet been, to the glory of God. Oh, my friend! we each of us alone know ourselves, the opportunities we have enjoyed, and the use we have made of them.

'Your last was penned when, with too much apparent reason, you almost despaired of our dear H. More's longer continuance amongst us. But Providence seems to intend graciously to prolong her invaluable life. In a better sense than that of the Roman poet we may say—

<blockquote>Præsenti tibi maturos largimur honores.</blockquote>

I rejoice in the testimonies of respect and gratitude which she is continually receiving from the United States, &c. &c.

'My dear wife has taken the pretty comfortable retirement from which I am now writing that I may more securely, 'from the loopholes of retreat,' peep at the great Babel that is near me, and at a safer distance hear the sound it sends from all its gates. It will render our retreat the more interesting to you to hear that it was the residence of the widow of good Bishop Horne. I thank God I am nearly restored to my ordinary standard of strength. This may find you on the other side of the Channel, for I remember you talked of a visit to Bellevue—a pleasure which I have myself not seldom enjoyed in prospect, though I never have been able to realise it. If you do visit the Emerald Isle, you must, I think, be tempted to cross the kingdom and pay your respects at Limerick. Have you read the Bishop's* speech? It is one of the most able ever delivered in Parliament, and I cannot but feel some indignation when I remember the coldness with which it was spoken of by many who ought to have felt its excellences with a keener relish. But they did not expect a debate, and were in haste to get away to their dinners. Incidents now and then occur which lead to conclusions much more unfavourable respecting our public men than they deserve. Our destination is uncertain. We may be ordered to Bath When do you settle again in your paradise? I beg my kind regards to Mrs. H., and am ever most truly yours, 'W. WILBERFORCE.'

* Right Rev. John Jebb, Bishop of Limerick.

The following extract from a letter* to him from myself is inserted on account of its special reference to the state of his health alluded to above, and also from its mention, after a lapse of twelve years, of the Latouche family, alluded to with so much interest in the first chapter of this work :—

'Bangor: October 18, 1824.

' My dear Friend,—At Milford, as we were on the point of embarking for Ireland, I was greeted by your most interesting and welcome letter. The animation which it breathed, and the firmness of the writing, confirmed the delightful assurance which it gave me, that you were restored to your ordinary standard of health; and oh! may it please God to vouchsafe you a long freedom from the painful attacks which have lately caused your friends so much anxiety. You were most kind in not deeming my suggestions about the expediency of allowing yourself a few years of comparative respite from public life troublesome. I well know that your vigorous mind reasons on the principle of

Nil actum reputans dum quid superesset agendum.

But two such severe illnesses as you have endured within a few months painfully demonstrate that the body is not equal to the demands of the mind, but

* Vide Correspondence, vol. ii. p. 482.

calls for relaxation from over-anxiety and oppressive care. One thing, I trust, you will be prevailed upon to determine, and that is very much to narrow the sphere of your labours, and not allow worthy but inconsiderate men to force too much work upon you.

'We paid a most interesting visit at Bellevue. Dear amiable Mr. Latouche bends beneath the weight of ninety-one years; but, considering his great age, he is a wonderful man, being in full possession of his faculties, and replete with attention to all around him. There is in his demeanour a benignity, a meekness, and a courtesy, which attract to him in return inexpressible tenderness and respect. He was generally the first at prayers in the chapel every morning. Of the chapel at Bellevue you have doubtless heard. It is connected with the house by a long glazed walk, planted on each side with exotics and flowering plants. It is built on an elegant design, and is well adapted to its object. The girls of Mrs. Latouche's school, which is in the park, attend the chapel regularly, and open the service by singing a psalm or hymn; and being carefully instructed for this purpose, their singing is truly beautiful, and imparts a peculiar interest to the family devotion. Mrs. Latouche is the very image of benevolence; she is ready at all times to make any sacrifice of time or attention for the good of her fellow-creatures.'

Under the date of Bath, October 27, 1824, I find the following note, which is merely inserted as an example of Mr. Wilberforce's very easy and playful mode of addressing his friends:—

'My dear Friend,—On coming down stairs after my siesta (for so I cover from myself the good plain English nap of an old gentleman in ancient days) I am agreeably welcomed by your letter. " But," says Mrs. Wilberforce, " why will not Mrs. Harford accompany Mr. H.?" "I am sure," I replied, "that it would add to our pleasure:" so immediately my lady proceeded to convey, by the enclosed note to Mrs. H. and you, our wishes. Hoping so soon to converse with you *viva voce*, I will now only add that I am ever sincerely yours,

'W. W.'

Soon after our return from Ireland we obeyed this kind summons from him to visit him at Bath, where he was taking the waters, and he and Mrs. W. spent about a fortnight with us in November, a period of high and peculiar enjoyment. From his state of health, and by order of his medical attendant, we asked but very few persons to meet him, and thus had the most uninterrupted enjoyment of his society. He needs no excitement from company. The playfulness and elasticity of his mind appear never to forsake him, and his wit and keen perception of humour render his conversation always enlivening. His anecdotes are inexhaustible and so well told! On one of these days, we drove to Clifton, and amongst other objects called at my brother's house, and had an interview with his children; the cordiality with which they welcomed Mr. and Mrs. W. in their father's absence, quite won their hearts. Mr. W. was peculiarly well, and in the most delightful spirits, charmed with the scenery, sometimes repeating poetry, sometimes reading aloud out of a miniature Cowper, sometimes singing. I half expected to see him take flight altogether, he seemed in such a winged state. He read with great delight Cowper's lines on the Squirrel, and said he thought it one of the happiest and most lively descriptions in any poet.

CHAPTER VII.

How Mr. W. was first induced to take up the Slave-trade Question—Granville Sharp—While staying at Holwood, Mr. Pitt urges him to bring it forward—He opens it in the House in 1789—Determined Opposition—Professor Sedgwick's Recollections of the brilliant Debates it called forth—Pitt's memorable Speech—Windham's Remarks on it—Cowper's Sonnet to Mr. W.—Mr. W.'s Recollections of Mr. Pitt—They visit France in 1783 accompanied by Mr. Eliot—Amusing Incidents—Mr. Pitt, Premier—Mr. W. becomes M.P. for the County of York —Pitt, Fox, Canning, Lord Harrowby.

I ONE day asked Mr. Wilberforce what had first induced him to take up the slave-trade question. He told me that Granville Sharp's philanthropic efforts on behalf of a runaway slave in 1780, &c., and a pamphlet of Clarkson's, had first turned his attention seriously to the subject, and that in the same year he had set on foot enquiries, through Mr. Gordon, a large West India proprietor, respecting the condition of the slaves in those colonies. 'Mr. Gordon was himself going out thither to inspect them; and I expressed my hope to him,' said Mr. Wilberforce, 'that the time would come when I should be able to do something in behalf of the slaves.' It was not, however, till the year 1787 that Mr. Wilberforce seriously resolved to take up the question. He had then acquired much information upon it, having often been amongst the African merchants, who had communicated to him many important particulars respecting the details of the traffic. 'In 1787,' said he, 'I was staying with Pitt at Holwood—one has often a local recollection of particular incidents—and I distinctly remember the very knoll upon which I was sitting, near Pitt and Grenville, when the former said to me, " Wilberforce, why don't you give notice of a motion on the subject of the slave trade? You have already taken great pains to collect evidence, and are therefore fully entitled to the credit which doing so will ensure you. Do not lose time, or the ground may be occupied by another." I did so, and upon that occasion Fox said he had himself seriously entertained the idea of bringing the subject before Parliament; but he was pleased to add that, it having got into so much better hands, he should not interfere. In 1788 I was in such indifferent health as to feel it to be my duty to entreat Pitt, in case it pleased God I should not recover, to promise to take up and prosecute this important object for me. He assured me that he would. In 1789 I opened the question to the House, and the line of argument which I pursued was this: I explained the nature of the traffic for slaves on the African coast—the flagrant evils, the wars, the cruelties, the barbarisms which it engendered—the obstacles which it opposed to the progress of civilisation—and also strongly dwelt on the horrors of the Middle Passage. Finally, I insisted on the impolicy of the trade, since by improving the condition and treatment of the slave population in our islands their numbers might be kept up by natural means. Sir William Dolbyn, and some other members to whom the subject was new at this time, hearing that a slave ship fitted out for the traffic was then lying in the river, went on board to examine the arrangements, and to test the correctness of some of my statements. They returned from the survey penetrated with indignation and horror, and added much to the impression which my speech had produced by what they reported. Yet there were persons who attempted to deny these horrors altogether, and spoke of the Middle Passage as a voyage of pleasure! Liverpool, it was asserted, would be ruined by the abolition. See how Liverpool has advanced in prosperity. The trade, it was argued, was essential as a nursery for our seamen. Clarkson proved to a demonstration the contrary. Burke, I remember, complimented me on my speech, and thanked me for the information he had received from it.'

The enmity and the calumnies prompted by self-interest which assailed Mr. W. in consequence of his taking up this question are scarcely to be imagined, and they were rendered still more bitter when it was found that nothing was able to shake his resolution to persevere. As an instance of the ridiculous stories told in consequence to his disadvantage, Clarkson was travelling in a stage-coach, when the conversation turning on the abolition question, one of the passengers gravely said, ' Mr. Wilberforce is no doubt a great philanthropist in public, but I happen to know a little of his private history, and can assure you that he is a cruel husband and even beats his wife.' At this time Mr. Wilberforce was a bachelor. Another current story was that he had married a black woman, and a third made it a lady's-maid. He was not a little amused while this report was running its round at receiving a letter from one of his political partisans in Yorkshire, who had lost caste by marrying his cook, congratulating him on his having surmounted the common prejudices of society, and cut out his own path to happiness.

The great and long-contested slave-trade battle will never lose its interest. Of some of the debates on that occasion Mr. Wilberforce remarked, ' Those were glorious nights in the House of Commons.' Wilberforce put forth all his powers; Fox and Burke were magnificent; but on one occasion in particular, Pitt, by common consent, bore away the palm of transcendent eloquence. Towards the close of this speech the great orator beholds through the vista of futurity the accomplishment of his brilliant anticipations for Africa. He beholds it civilised, free, and happy—illumined by the beams of science and by the brighter beams of true religion.

He was in the grand movement of his peroration, and was contrasting the glories of England as the land of science and art and political freedom, with what it was in its days of savage darkness, when the sun rose and sent his horizontal beams through the windows of St. Stephen's chapel. His imagination kindled at the sight, and he broke out, as if in a spirit of prophecy, into that beautiful passage from Virgil's Georgics, i. 250-1 :—

> ——nos primus equis oriens afflavit anhelis;
> Illic sera rubens accendit lumina Vesper.

And in anticipation of a brighter future for Africa he continued : ' Then, sir, may be applied to Africa those words, originally used indeed with a different view :—

> His demum exactis ——————
> Devenere locos lætos, et amœna vireta
> Fortunatorum nemorum, sedesque beatas :
> Largior hic campos Æther, et limine vestit
> Purpureo.'

On this subject Professor Sedgwick, a high authority, writes as follows :—' I am old enough to remember the dreary time when the brave indignant oratory of Fox, the majestic eloquence of Pitt, and the silver voice of Wilberforce (speaking like an angel in the cause of mercy and truth and national honour), were heard in vain in St. Stephen's chapel; when, year after year, the representatives of free England sanctioned and commended a vile unchristian trade in the flesh and blood of the men of Africa. Vain were the pleadings of Christian love and national honour, when the children of Mammon were allowed to hold the balance while the debate was going on.'*

Cowper's beautiful sonnet, on the same subject, will fitly follow in the train of the above statements:—

TO WILLIAM WILBERFORCE, ESQ. 1792.

Thy country, Wilberforce, with just disdain,
Hears thee, by cruel men and impious, called
Fanatic, for thy zeal to loose the enthrall'd
From exile, public sale, and slavery's chain.
Friend of the poor, the wronged, the fetter-gall'd,
Fear not, lest labour such as thine be vain!
Thou hast achieved a part, hast gained the ear
Of Britain's Senate to thy glorious cause:
Hope smiles, joy springs, and tho' cold caution pause
And weave delay, the better hour is near
That shall remunerate thy toils severe
By peace for Afric, fenced with British laws.
Enjoy what thou hast won, esteem and love
From all the just on earth and all the bless'd above!

On the breaking up of the House, after the memorable debate in which Pitt spoke so finely, Windham, who was by no means partial to Pitt, and who did not take the abolition side, met Wilberforce and accosted him thus—'Really, if your friend Pitt should speak often as he did last night, he will make converts of us all. It was as if he were inspired.'

Having mentioned this great statesman thus particularly, I will here introduce various memoranda respecting him communicated to me on different

---

* See Prefatory Letter, p. 35, to Dr. Livingstone's Cambridge Lectures, by the Rev. Professor Sedgwick, M.A. F.R.S. &c.

occasions by Mr. Wilberforce, who was fond of referring to his illustrious friend. Their acquaintance first commenced as under-graduates at Cambridge; 'but,' said Mr. Wilberforce, 'we were not intimate there.' After quitting the University they often met in the gallery of the House of Commons, and both entered Parliament when just of age, in the same year, 1780. From this time their acquaintance grew into close intimacy. Pitt lived much in what may be termed a select club of his personal friends, young men of great talent, most of whom looked to him as their political leader. They were about twenty-five in number, and met at the house of a man named Goostree, in Pall Mall. Among them were Pratt (afterwards Lord Camden), Althorp (afterwards Lord Spencer), Grenville (afterwards Lord Grenville), Robinson (afterwards Lord Rokeby), Smith (afterwards Lord Carrington), Lords Duncannon, Euston, &c.; Mr. Pitt, Mr. Bankes, Mr. Wilberforce, &c. &c. Mr. Pitt had the reputation in after life of being remarkably distant and reserved in manner, and he was apt to be so amongst strangers. 'I remember,' said Wilberforce, 'being present at a dinner when Mr. Ryder, afterwards Lord Harrowby, was one of the guests, and was first introduced to Pitt, with whom he subsequently became officially connected and lived in terms of friendship. Pitt was very silent throughout the afternoon, and I knew that Ryder's presence was the real cause of it; but when amongst his friends he was one of the most agreeable of men, full of wit, playfulness, and vivacity. Bankes and I were already in possession of our fortunes: therefore his house in town, and my villa at Wimbledon, were the scenes of our frequent meetings.

'Pitt's delight was to sleep in the country, and therefore,' said Mr. Wilberforce, 'he and I, when the House did not sit very late, were in the habit of driving to Wimbledon together; and on one occasion he lived thus with me for months. Often and often have I called him in the morning, and the capacities of the villa were frequently put to the utmost stretch to accommodate our various friends.

'I was present when Pitt delivered his first speech in Parliament, in support of Burke's motion for economical reform, and it was so perfect in its way that he can scarcely be said to have surpassed it in the full maturity of his powers. When, at the early age of twenty-three, he took office as Chancellor of the Exchequer, under Lord Shelburne (1782), the difficult task quickly fell upon him of coping, almost single-handed, with a majority of the House of Commons, led on by Fox, Burke, Sheridan, Lord North, &c. His great mind rose to the full level of the occasion, and he soon proved himself to be in no way inferior to his illustrious father in the finest qualities of an orator and a statesman.

'In particular, when he undertook the difficult task
L

of defending the recent Peace, on February 21, 1783, I well remember that Pitt was so oppressed by a severe sick-headache as to be scarcely able to hold up his head. Fox assailed him in a very able speech, in the midst of which Pitt was obliged, from actual sickness, to retire to the entry-door, called "Solomon's Porch," behind the Speaker's chair. I seem to see him holding the door in one hand, while he yielded to his malady, and turning his ear towards the House, that if possible he might not lose a single sentence that Fox was uttering. Never do I recollect to have witnessed such a triumph of mind over physical depression. When Fox sat down he replied to him with great ability, though with less brilliancy than usual; but, on a renewal of the same discussion a few days after, in a different form, he made one of the finest speeches ever delivered in Parliament.'

In the autumn of 1783, after the breaking up of Lord Shelburne's Cabinet, Mr. Pitt spent some time in France, chiefly with the view of acquiring the habit of speaking French fluently: Mr. Eliot—who afterwards married his sister—and Mr. Wilberforce were his companions. Before setting off, they met at the seat of their friend, Mr. Bankes, in Dorsetshire. Speaking of this visit to me in 1825, Mr. Wilberforce smilingly said : ' Ah, it was at Bankes's that I was near shooting Pitt, poor fellow ! · At least it was a standing joke amongst our mutual friends to lay this to my charge. He was passing through a hedge, and I, not knowing it, fired in that direction; but he was not in any real danger, for none of the shot passed near him. It is now forty-three years ago.' They fixed themselves first at Rheims, and owing to their having quitted England in a hurry they brought no introductory letters with them, excepting one procured for them by Mr. Thelluson, afterwards Lord Rendlesham. It led to an amusing incident, which Mr. Wilberforce often related with much humour. The letter was sent to the person to whom it was addressed at Rheims upon their arrival. The next day he called. He was a smart-looking man, with a bag-wig and sword, and his manners were easy and polite. After conversing with them for some time, and offering his best services, he took his leave. They quickly returned his call, and expected from the appearance of their new acquaintance to have found him living in one of the best mansions of the town; but to their no small surprise and amusement they found themselves at the door of a little grocer's shop, where behind the counter stood their friend with his apron on, doling out pennyworths to his customers. Not at all abashed, in an instant he doffed his apron, showed them into an inner parlour, and talked away with French ease and vivacity. He frankly told them, on their enquiring whether he had any acquaintances amongst the gentry of the city, that he had none; but he said he would go to the Commandant, and apprise him of their arrival. He did so, and the intelligence was soon after communicated to the Archbishop, who, on hearing that three Englishmen, members of Parliament, and one of them calling himself Mr. Pitt, were in Rheims, was incredulous as to the fact. However, he sent a Monsieur de Lageard (who had been in England to give evidence on the Douglas cause) to call and report the result. This gentleman quickly found that all was right, and was so pleased with his new friends that he spent some hours in their company. After this the Archbishop paid them great attention, as did also the principal families of the place. The story of their *début* at Rheims soon got wind, and it so amused Marie Antoinette, that after they were introduced at Court she seldom failed, on seeing Mr. Pitt at her parties, to enquire whether he had heard from his friend the épicier.*

Another droll anecdote connected with their stay at Rheims was as follows:—Supping one evening at

---

* In some slight particulars the above account of the visit to Rheims differs from a letter in which he details them to Mr. Bankes, in 1783; but as Mr. W. on two or three occasions related the facts to me as I have told them, I can only account for the variation by the effect of the lapse of time on his memory.

the house of one of the gentry they observed a large space left vacant in the middle of the table, which was in other parts covered with delicacies. The party was large, and a mysterious smile and whisper circulated amongst the guests. Presently the door of the apartment flew open, and two servants entered bearing between them a huge dish of roast beef, which was put down with an air of importance in the vacant place. The beef was cut, and the Englishmen invited to fall to. It proved to be almost raw, and they could none of them touch it. As to Mr. Pitt, he never ate any meat which was not thoroughly well done. The French appeared much disappointed, and it was clear they thought their guests were ashamed to display their national tastes before strangers.

From Rheims they proceeded to Paris, and thence to Fontainebleau, where the court then was. Here they received great attention. Marie Antoinette shone amidst the brilliant assemblage—that star of grace and beauty described by Mr. Burke with such chivalrous feeling. Though Mr. Pitt was shy of talking French, he expressed himself when he did converse with great correctness, and left behind him some impression of his wonderful powers. Mr. Wilberforce often mentioned as a proof of our national prejudice, or *gaucherie*, the determined manner in which the English hung together at the French court. Towards supper-time the brilliant mass was in the habit of breaking into parties and supping in sets together. Instead of mixing with the agreeable foreigners around them, the delight of the English was to have a table to themselves; and they admitted no Frenchman but the Marquis de Noailles, who spoke English fluently. Even the British Ambassador, the Duke of Manchester, was to be found amongst these exclusives.

Soon after they returned home, Mr. Pitt became Premier, and Mr. Wilberforce one of the representatives of the county of York. As their intimacy was of the most confidential description, few men have enjoyed such opportunities of judging of the qualities of another, both of head and heart, as Wilberforce had of Pitt. 'Very few people,' said he one day to me, 'understood Pitt, he was so shy. He was a truly kind-hearted fellow. His feelings were so tender, that he could not endure to hear of an act of cruelty, yet in matters of principle his firmness of purpose was inflexible and his courage undaunted. There was in him a moral elevation and greatness of soul which raised him far above the level of most of his adherents. He truly loved his country, and sought its good. I feel sure his love for it was such that he would readily have consented any day to die for it, even though he were aware that it would never be known that he had made such a sacrifice. I never saw a man more entirely free from vanity. In this respect he was very unlike Fox, who had a great share of it.

'His memory was wonderfully accurate and retentive, and when he indulged in a classical quotation in his speeches it was always most happily introduced. When Fox was speaking, Pitt usually made notes of what fell from him; but on rising to reply I never saw him refer to those notes. He excelled every person I have ever known in the faculty of steadily fixing before the eye of his mind all the parts of a complex question, and of accurately weighing their respective bearings, giving due force to each in forming his conclusion. A beautiful part of his character was his perfect fairness in argument. When you reasoned with him he listened with patience to all that could be urged against his opinion, and never manifested a disposition to undervalue a single word that you urged in opposition to him. Mr. Pitt was the wittiest man I have ever known, but his wit was of a very peculiar cast. Wit in most men consists in brilliant flashes of fancy, or in suddenly striking out in the heat of conversation unexpected coincidences between dissimilar objects and ideas. In Pitt it appeared to be a pure operation of the intellect. It seemed as though the forms of all objects were so present to the view of his comprehensive and cultivated mind that he could combine, oppose, or compare them in such a manner as to excite at his will unlooked-for coincidences, surprise, and pleasure. His powers of repartee were very great, but always under the restraint of good humour. His love of truth was remarkable. I have known instances of it in cases where nothing but the motive of high and unbending principle could have ensured his tenacious adherence to it. His failing as a wise man was an over-sanguine estimate of the chances of success under contingent circumstances—disposing him to believe what he wished—and in the case of foreigners too easily to confide in statements on which it was important for him, as a statesman, to obtain the most accurate information. His great ambition was to be a Peace-Minister. The French war he considered as forced upon him, and it deserves to be recorded as an instance of the short-sightedness of the wisest statesman that on entering upon it he said it would be over in a twelvemonth. Burke took a different view. He, Pitt, Dundas, and myself were together one day just before the Revolutionary War broke out. Dundas said there is no avoiding war: to war we must go, but it will only be a short bout of it. Pitt replied nothing, but Burke said: "Mr. Dundas, I fear you make a very mistaken estimate of the probable duration of the war, which I am persuaded will not only be long, but will require England to put forth all her energies in the course of it." I remember another mistaken anticipation of his. It seems to me but yesterday that he said in my hearing—Windham, I recollect, was standing near him at the time—that although it might be presumption in him to point out the very day on which it would be impossible for the French Government (beggared as he knew them to be in their resources) to go on, yet he would almost venture to name the week. He made this remarkable statement only just a fortnight before the battle of Marengo.

'He came into Parliament a finished orator. We all expected great things from him. He had belonged to a club of lawyers in town, and had occasionally given proofs at their meetings of his great powers. I had not heard him. But we were none of us prepared for the extraordinary superiority in debate which he manifested from the first. I might almost say his first speeches were equal to his last. I do not, of course, mean in comprehension, or in those qualities which are the results of mature experience; but in those which constitute an accomplished and powerful orator. I have sometimes thought that Canning, in his finest speeches, reached the elevation and dignity of Pitt, while it is unquestionable that his powers as a speaker were more diversified—yet there was always this great difference: you never forgot it was Canning while he was speaking, but Pitt often made you forget the orator in his subject, and hurried you along with the full tide of his majestic eloquence. I am old enough to recollect some few members of the House who were amongst his father's contemporaries, and they agreed in pronouncing his son to be his superior as an orator. The style of Lord Chatham's speaking was, however, very different. It was in its ordinary fabric conversational—then it would rise into lofty bursts of eloquence; but much of its effect is to be attributed to that astonishing degree of personal ascendency which he had acquired, and which at times positively made the House quail, as it were, before him. There was a great deal of the theatrical in Lord Chatham's character. The fault of Pitt's speaking was its being too uniformly dignified and stately. It wanted breaks to give full effect to the finer passages. On the whole, I would say, the intellectual powers of Pitt were superior to those of any other man I have known.

'Though Pitt was, as I have stated, so fine a classical scholar, he had no time, coming into office as he did so early in life, to pursue such studies. I remember having a point of business to discuss with him, upon which, before he came to a decision, it was necessary to consult Lord Grenville. He had put it off again and again, but at length fixed a day on which he was to dine with Grenville, when he assured me it should not be forgotten. I went to him the next morning, when he said: "Wilberforce, I am really ashamed to see you, for I have not done your business; and to tell you the truth, we did not even touch upon the topic—for it so happened that, after dinner, we got talking upon a passage in Cicero. One of us then alluded to a passage in Homer upon which we could not quite agree, and Grenville went to his library to bring the volume which contained it. After this we got so deep into classical discussion that, on looking at our watches, we were not a little surprised to find it was past eleven. It was then too late for your business."'

The late Lord Harrowby told me that he was staying at the house of a friend, where Mr. Pitt was one of the party, when it so happened that himself and two other gentlemen had not been able to agree upon the meaning of a dark passage in Thucydides. While they were discussing it, Mr. Pitt entered the room, and they told him what was going on between them. He took up the book, and reading the passage in the Greek, immediately, without hesitation, translated it into English in the most definite and elegant manner. Let me here add that Mr. Wilberforce more than once spoke to me in the highest terms of Lord Harrowby as a statesman, saying that, if his health had been equal to his intellect, he was the person of all others who ought to have been at the head of the Ministry with which he was ultimately connected. He had, he went on to say, beyond most men whom he had known, the faculty of so surveying any complex and difficult question as to be able to keep in view and to compare all its various parts and relations, and thus to bring an enlightened judgment fully to bear on his final decision.

'Pitt was a true lover of the country and of the beauties of nature, though of course he had very little leisure to indulge this taste. At Holwood he had a coppice wood of about thirty or forty acres, with many fine timber trees scattered throughout it; and he showed his skill in landscape gardening by carrying paths through it in such a way as to bring the finest trees into full view. Often have I seen him and Lord Grenville at work in this wood, each with an axe in his hand, as happy and busy as possible. It was much to Grenville's honour that he never flattered Pitt, which I cannot say of many others of his associates. They disagreed in their opinions upon Parliamentary Reform, of which Grenville was a decided opponent.

'From his youth he was delicate in health, and frequently suffered from suppressed gout. The languor incident to this complaint, and the exhaustion of business, frequently led him to take too much wine. He died of complete exhaustion; in fact, his stomach was gone for some time before his death.'

Mr. Wilberforce to his latest day seldom mentioned Mr. Pitt's name without some affectionate epithet, and he once said to me : 'I certainly never knew, on the whole, so extraordinary a man.' Occasionally, when thus speaking of him, I have heard him express his deep regret that, owing to Mr. Pitt being so entirely absorbed in politics, he had never allowed himself time fairly to turn his attention to religion, or to examine Scripture as the rule of life.

He expressed a similar regret with respect to others of his contemporaries with whom he had mingled much in public life, lamenting, not only in their case, but in that also of many learned theologians, the want of a more experimental acquaintance with the Holy Scriptures, adding : ' I question much whether a large proportion of such men be not very deficient in that intimate knowledge of the Bible which we ought all to cultivate. Many of them may be able to give a good account of disputed passages, or talk well as critics and grammarians; but how different is this from a heartfelt delight in the Bible. They know not the Scriptures as a man knows his friend—as he almost knows by the sound of his voice and the expression of his face what he is about to say. They have not that knowledge of it which transmutes the heart and character, by an assimilating influence, into the image of the principles which it unfolds.'

CHAPTER VIII.

1825 TO 1828.

Retirement of Mr. W. from Parliament—The Author visits Staffa and Iona—Honourable Mention of Mr. W. there—A Visit from Mr. W. in December 1825—His Remarks on Castlereagh—Sheridan—Dr. Johnson—Burke—Various Anecdotes of Distinguished Men—Dr. Carey, the Baptist Missionary—Eloquence and Wit of Canning—Lord Brougham—A Visit to Mr. W. in Bath in 1827—Mr. W. visits Yorkshire—Attentions paid him there—In 1828 Visits to him at Bath, &c.—Mr. W.'s Seventieth Birthday.

EARLY in the year 1825, the following letter reached us from Mrs. Wilberforce :—

'Uxbridge Common: Feb. 5, 1825.

'My dear Mrs. Harford,—I am allowed to communicate to you and Mr. Harford what is at present a secret, for Mr. Wilberforce wishes it not to be talked of for a day or two. In consequence of an interview with Dr. Chambers, and a visit from his friend, Mr. Babington, and his own reflections on all that has passed, he has made up his mind to vacate his seat in Parliament immediately, and to retire from public life to quieter scenes, to which his health and strength are now more suited. I have yet only communicated the intelligence to my mother, and my dear sister will have it from me by this day's post. Knowing your and Mr. Harford's kindness, and your feelings towards him, I wished immediately to tell you the joyful tidings. Those amongst his friends who have not, like yourselves and Mr. Babington, seen him in private, since his late alarming attacks, or heard of Dr. Chambers's opinion, may possibly doubt whether he has done right in taking this step; but others will, I hope, cheer him with their approbation. We are busily engaged in seeking a residence suitable to him for the remainder of his days. I hope when you and Mr. Harford visit these parts you will not forget us.

'I am glad I can now join with you in wishing for the protraction of the present Parliament. While I considered my husband's life in danger from its continuance you will not wonder that I hailed with gladness every report that its close was nigh.

'I rejoice to say that my "gude man" is wonderfully well for him at present, and his life here suits him à merveille. With kind and affectionate regards to you and Mr. Harford,

'I am, &c. &c.

'B. W.'

We were quite prepared for the intelligence conveyed in this letter; and though it was impossible not to regret that the accents of that voice which had so long, so energetically, and so eloquently pleaded in the British Senate the claims of suffering humanity, and the highest and best interests of his country, would never again be heard within its walls; and though one could not stifle sentiments of keen regret at the irreparable loss which Parliament would sustain by the retirement of a man of such great talent and of such incorruptible principle, yet this reflection was checked by the conviction that he had already run his allotted course, and finished his high Providential destiny, and that it now only remained for him to pass the evening of his days in tranquil blessedness. Yet it was justly hoped, and so it proved, that in a more private sphere his influence would still exert itself beneficially in numberless ways, and especially in promoting the final triumph in all its parts of that great cause to which he had more especially dedicated his life. Our predominant feeling, therefore, was thankfulness, from the hope that, by his withdrawing from a scene to the fatigues of which he was no longer equal, his valued life might be for many years preserved to his family, to his friends, and to society.

On the very day that his writ was to be moved for, I received the following few lines from him, under the last of his franks to me, succeeded by a letter some months after, in which he more fully developes his feelings on this occasion :—

'Near Uxbridge: Feb. 22, 1825.

'My dear Friend,—I do not now take up my pen to answer your last truly friendly, and to me deeply interesting letter, because dinner is actually on the table; but a word that dropped from you intimated that you wished to hear from me again before I should have put off my parliamentary character, and therefore as my writ, I am told, is to be moved for this day, I take up my pen just to say *Vale*, and to promise in some early season of leisure to write you a letter less unworthy of the name. Meanwhile this *letterling* will not be without its value (*inest suɑ gratia parvis*), since it shall assure you that I look back with no small pleasure on the happy days we lately spent under your friendly roof, and that, looking forward with hope for the renewal of our intercourse, I am, with kindest remembrances to your dear lady, sincerely and affectionately yours,

'W. W.'

The active mind of Mr. Wilberforce, as above anticipated, though released from the shackles of public life, was still occupied by an earnest desire to be useful to his fellow-creatures in a quieter and less conspicuous way; and, amongst other projects, he intended that his pen should have been subservient to this purpose. On trial, however, he found the needful application injurious to his health. The following letter forms a comment on these remarks:—

'Elmdon House: Oct. 6, 1825.

'My dear Friend,—Although I have been for some time in circumstances which leave little leisure for letter writing (having been paying a succession of short visits to a number of old and dear friends), the want of time alone should not have prevented my writing to you. But it is only during that very period of the day which is usually devoted to air and exercise that I can use my pen, for my eyes require absolute rest for many hours after rising in the morning. But I am always tempted to be diffuse on this head when I am writing to my friends, from feeling as though I had to vindicate myself; but you, I trust, will give me credit for all friendly sentiments and feelings wherever you are concerned. Your lively description of the beauties and sublimities* of Scotland create in me a strong disposition to know them *personally*, if I may use the phrase. But I almost shrink from undertaking any tour which would require the expenditure of many weeks. To

* I had sent him a particular detail of the scenery of Staffa.

you I open my heart, and I will therefore confess that the consciousness of my not having duly used the very long life (very long compared with what had been anticipated by medical men) which Providence has already granted, renders me very desirous of not wasting what may yet remain to me. But, alas! with these impressions deeply imprinted on my heart, week after week has been passing away in what too much resembles Shakespeare's expressive phrase of "shapeless indolence." It is true I have been for a considerable part of the period in circumstances very unfavourable to study, more especially to composing. Before I can write anything that at all deserves to be read, I must exercise my mind much; for I am a very slow composer, and just now I am called off from the subject that was about (as I hoped) to occupy my thoughts. The fact is, I am deeply impressed with a sense of the bad consequences of suffering our artisans, &c., to remain wholly ignorant of the evidences of Christianity, while they are to be well instructed in the various branches of philosophy; for I fear that such a procedure will tend but too surely to make a set of self-conceited sceptics. Do think over the subject; and I do not say write me your opinions, because I look forward with pleasure to the prospect of seeing you and Mrs. H. once more, though we have been compelled to be too late in our visit to Bath. But the attraction to Blaise Castle will be too strong to be resisted. Farewell! With our kindest remembrances to Mrs. H.

'I am ever affectionately,

'W. W.'

And here, as allusion is made to my descriptions of Scotland, I must mention that we visited Staffa and Iona in company with his old and much-valued friend, Mr. Babington * (whom we unexpectedly met at Oban), who had shared with him in all his schemes of benevolence, both in and out of Parliament. On our entering the school-room at Iona, the schoolmaster, hearing the name of Babington, enquired whether it was the same gentleman who, together with Mr. Wilberforce and some others, had done so much for the promotion of Christian education and instruction in that remote region. Finding that it was so, he expressed his grateful sense of their united benevolence, and said that the school-room, built and supported at their expense, had also been used for public worship, in the absence of any chapel, by means of ministers who periodically crossed from the mainland for that purpose. I thought of Johnson's fine sentences in reference to Ioua in his tour to the Hebrides, and felt that the excellent men who had

* M.P. for Leicestershire during many years.

thus befriended the highest and best interests of its inhabitants would have won the praise of the great moralist, had he visited Iona at the time that I myself trod its shores. I observed the sparkle of benevolent delight in Mr. Babington's eye at this occurrence.

Towards the latter end of October Mr. Wilberforce arrived in Bath; and having some friends at that time staying with us, whom I thought it would be a great pleasure to him and Mrs. W. to meet, I pressed him to give us a few days in anticipation of the longer visit which he had promised us when his course of Bath waters should be concluded.

This visit did not take place; but we had the pleasure of receiving them in December.

During his stay with us his conversational powers were constantly exerted for our delight and amusement, often on religious subjects of the highest import and interest, and at others in recurrence to his recollections as a statesman of the busy scenes in which he had been so long engaged. He sometimes insensibly slid into critical remarks on the style and manners of contemporaneous statesmen, telling anecdotes respecting them.

Of Lord Castlereagh as a speaker, Mr. Wilberforce said : 'When he was in his ordinary mood he was very tiresome—so slow and heavy—his sentences often only half-formed—his manner so confused, like what is said of the French army in the Moscow retreat, when horse, foot, and carriages of all sorts were huddled together, helter-skelter; yet, when he was thoroughly warmed and called forth, he was often very fine, very statesman-like, and seemed to rise quite into another man.' Of Sheridan's famous speech on the Begum question he remarked: 'It was a most surprising exhibition—five hours; and yet we were none of us tired. What a compliment was paid him by Pitt when, at the conclusion of his speech, he proposed that the House should adjourn, in order that we might calmly survey the question on which we were to vote in a state of freedom from the spell of the enchanter by whom we had been so fascinated. Our general impression always was that he came to the House with his flashes ready and prepared to let off. He avoided encountering Pitt in unforeseen debating; but when forced to it usually came off well. I remember the sort of theatrical effect he gave to his philippic against Hastings, whom he described as one moment guilty of State crimes on an enormous scale, and at another stooping to most petty meannesses—with one hand grasping a sceptre, and with the other picking a pocket. How caustic was he upon Dundas, when, in one of his replies to him, he said: " The right honourable gentleman has applied to his imagination for his facts, and to his memory for his wit."

'Sheridan was a jolly companion, and told good stories; but has been overrated as a wit by Moore.

'Pitt talked in a lively way amongst his friends. Fox in general society was quiet and unassuming; but as an orator he was often truly wonderful. He would begin at full tear, and roll on for four hours together, without tiring either himself or us.'

He spoke with high admiration of the great and various powers displayed by the Duke of Wellington, and of the electrical effect which a few words only from his lips produced upon his army when on the verge of battle. 'Sir Gregory Way described to us most picturesquely,' said Mr. Wilberforce, 'the particulars of an attack of the French on part of our army in the course of the Spanish war. "Our own people," said he, "were upon an eminence. Wellington was there. The French made up their minds to attack. A column of twelve or fourteen thousand men—their music playing, and their arms glittering in the sun—approached and pushed up the eminence. It was one of the grandest sights possible to a soldier's eye. Wellington allowed them to approach so near that we felt uneasy—we all watched his eye. He just then called out, 'Come, can't you give them a touch?' Our men rushed forwards—they had the advantage of the ground. The French were broken in a moment. Our attack was like a stream of lava from the mountains, overwhelming all that opposed it."'

Mr. Wilberforce delighted in music, and often took part at the close of a social evening with others in singing or chanting. As we were one day talking upon the subject of devotional poetry he said: 'Dr. Johnson has passed a very sweeping condemnation upon it, and has given it as his opinion that success in this species of composition is next to impossible; and the reason which he gives for it is, "that all poetry implies exaggeration; but that the objects of religion are so great in themselves, as to be incapable of augmentation." One would think, however,' said Mr. Wilberforce, 'that religion ought to be the very region of poetry. It relates to subjects which above all others agitate the hopes and fears of mankind—it embodies everything that can melt by its tenderness or elevate by its sublimity—and it has a natural tendency to call forth in the highest degree feelings of gratitude and thankfulness for inestimable mercies. His prejudice, poor man! in this respect, appears to me to resolve itself into the same cause which prevented his deriving comfort from the cultivation of religion. The view which he took of Christianity acted more strongly upon his fears and his superstitions than upon his affections. It inspired him with terror;

and, therefore, it often failed duly to influence his conduct, or to impart comfort to his feelings.'

Referring to the beauty of English villas he remarked: 'Well, I must speak of the security and comfort of English cottages. It is delightful to think how many there are in this country who, though having no title to personal security from the extent or importance of their possessions, are so completely guarded in their little nooks and tenements by the power of the law, that they can enjoy undisturbed every comfort of life as securely as the first peer in the land. I delight to see, as one sometimes sees, an old worn-out sailor, poor fellow! seated in his queer boat-like summer-house, smoking his pipe, and enjoying himself in a state of the most happy independence.'

Speaking of the caprice of many of the envied possessors of beautiful places, 'I remember,' said he, ' once dining with the Duke of Queensberry, at Richmond, with a small and very select party. George Selwyn, Pitt, Lord and Lady Chatham, and the Duchess of Gordon, were among the guests. We dined early, that some of them who were going to the Opera in the evening might be in time. The dinner was sumptuous—the views from the villa looked quite enchanting. The Thames was resplendent; but the Duke looked on with indifference. " I am quite tired," he exclaimed, " of the Thames. There it goes —on, on, on—always the same." '

We were talking of the levity of the French, and of their gaiety of heart, even under the severest misfortunes. This subject drew from Mr. Wilberforce the following anecdote, which he said had been related to him by Mr. Pitt :—' Shortly after the tragical fate of Marie Antoinette, a French *émigré* of some distinction, a denizen of the French Court, who had become acquainted with Mr. Pitt by meeting him there, took refuge in England, and on coming to London went to pay his respects in Downing Street. As was natural the conversation at once turned on the terrible scenes of blood which had recently occurred in Paris, and in particular on the cruelties wreaked upon the Queen. The Frenchman was at length quite overcome, and he sobbed out, " Ah! Monsieur Pitt, la pauvre Reine, la pauvre Reine!" These words were pronounced with much emotion, when a new idea possessed him, and starting from his chair he exclaimed, " Cependant, Monsieur Pitt, il faut vous faire voir mon petit chien danser." Then pulling a small kit from his pocket, and calling out, " Fanchon, Fanchon, dansez, dansez! ' he and the dog began to cut such capers together that the Minister's gravity was quite overcome, and he burst into a loud laugh, scarcely knowing whether to be most amused or astonished.'

As a further illustration of French vivacity, he mentioned the following anecdote of Bourdaloue, the celebrated preacher: 'Being appointed to preach on a special occasion before Louis XIV. and his Court at Versailles he did not make his appearance at the appointed time. This, of course, excited much surprise, and messengers were sent to his apartment to apprise him that the King was waiting. They knocked in vain, and, on looking through a chink, saw the Bishop skipping about his room to the sound of a kit which he held in his hand. "Monseigneur," as soon as entrance was obtained they exclaimed, "Monseigneur! are you aware that the King is waiting for you?" "Is it possible!" he replied: "to tell you the truth, I found myself so exhausted by long fasting that I felt unequal to preach, *sans faire cela, pour me réjouir un peu.*"'

The habit of hoarding money being touched upon by one of the party, he was asked whether he remembered Elwes, the miser, in Parliament. 'Yes,' he replied; 'and his name is connected in my mind with a most amusing incident. He wore a wig—I doubt whether he ever bought one—it looked as if it might have been picked off a hedge or a scarecrow. At that time we used to wear dress-swords occasionally at the House; for instance, if going to the Opera. One day, Bankes, whose carriage is stiff and lofty, had on his sword, and was seated next to

Elwes, who leant his head forward just as Bankes was rising up to leave his place, when the hilt of his sword came in contact with Elwes's wig, which it whisked off and bore away. The House was instantly in a roar of laughter. I never shall forget the scene. There was old Elwes, without his wig, darting forward to reclaim it; and Bankes marching on quite unconscious of the sword-knot which he wore, and wondering what the laugh was about.'

One day Mr. Wilberforce interested us much by the following remarkable fact:—'Dr. Colthurst, of Cambridge, was a very religious man, at a time when it was no light cross to support that character there. A poor man who had committed a capital crime had been tried and convicted. The Doctor attended him in prison—a species of charity scarcely then heard of amongst gownsmen—and it pleased God to bless his pious labours so much that the man's mind became deeply impressed; and, as the day of his execution drew near, he earnestly importuned his kind and pious instructor to attend him to the gallows. The Doctor at first shrunk from this painful office, and said it was one which he could not possibly, as a gownsman, undertake; but the poor man's entreaties were so earnest that he at length gave way and attended him to the last. His conduct was much talked about and discussed at Cambridge, and severely censured by many as fanatical and extravagant. In those days it was the custom at Cambridge for the gownsmen to frequent a particular coffee-house, where there were forty or fifty little tables, all of which were often occupied. Dr. Colthurst came in at a time when this was the case, and quietly seated himself at a table at one end of the room. At the other end was seated a Mr. ——, a Fellow of King's, a profane thoughtless creature, and who loved his joke. He took his opportunity, and called out, so as to fix general attention: "How are you, Dr. Colthurst? So I find you attended *your friend* to the last. Did he leave any message with you?" "It's true," replied the other, "I did attend him to the last." "Well," rejoined ——, "I admire your fidelity to *your friend.*" "Mr. ——," replied the Doctor, "should you ever unfortunately be in a similar situation, I'll be ready to do the same for you. Depend upon it, I'll attend you to the last." This rejoinder, uttered in a quiet good-humoured way, turned the laugh against ——, and put him down. This same person subsequently went into *orders* at fifty to take a living. It may easily be imagined what sort of clergyman he made. A circumstance, however, occurred soon after which was very remarkable. He one night dreamt that our Saviour appeared to him and said, in a solemn way: "What account canst thou render to Me of the sheep committed to thy care?" He woke, most painfully agitated by his dream; nor could he ever shake off the painful impression which it left behind. But though his nerves were shaken his heart did not appear to be changed—he was still the same man. He died shortly after, melancholy and dejected.'

He mentioned a curious encounter which one day took place between Dunning and Lord Mansfield:— 'Lord Mansfield once interrupted Dunning when pleading before him with the exclamation, "You are quite mistaken, Dunning." Dunning: "No, my lord, I am not."—Lord M.: "Then I had better go home and burn all my law books." D.: "No, my lord, you had better go home and read them."'

Speaking of Lord Chancellor Eldon, he passed a high encomium upon his judicial integrity, and said: 'I feel sure he would rather die than make an unjust decision.' Then, reverting to his recollections of him as Sir John Scott: 'His physical powers are really wonderful. I have known him, when a debate in the House of Commons has been protracted as late as eight in the morning, go down from the House to the courts of law to pursue his arduous duties as Attorney-General, without retiring to take any repose.'

The story is well known of the remarkable way in which Lord Eldon's great legal talents were first called out into public notice. George Harding, a barrister of eminence, had been retained in an electioneering cause, and just as the day approached for the pleadings he was taken ill and could not proceed. His clients were greatly alarmed, and urged him at least to point out some substitute of sufficient industry and ability to make himself master of the case in the few hours which were to intervene before the cause came on. He recommended John Scott, living in an obscure chamber of Lincoln's Inn, and a name at that time almost unknown. Scott, on being applied to, hesitated whether he could possibly undertake so difficult and complicated a case at such very short notice. As the story goes, added Mr. Wilberforce, ' he just then heard one of his children cry, which determined him—but this is doubtful. He did wonders in this cause, and from that day his fame was established.'

Talking of the habits of London thieves, he told us how his friend, Mr. Morton Pitt, who had been robbed of some plate, and who had in consequence been among the Bow Street officers, who were employed in enquiring after it, became curious, from what fell from them, to visit one of the night cellars to which the most notorious resorted. Accordingly, he was taken by a Bow Street officer in their confidence to one of these receptacles, who said, on introducing him : 'This is a person you may depend upon.' There was a supper going on, and the party assembled was very merry and joyous. They all professed great loyalty to the King, and talked much of the Royal family. In particular, they commiserated the hard fate of the Princesses in being condemned to a single life or an unhappy marriage. The man next whom Morton Pitt sat startled him not a little by suddenly enquiring, as of a brother rogue, 'which of the judges would you most like to be tried by?' 'Why, really,' he replied, 'the question has never happened to cross my mind.' 'Indeed!' said the other, 'I'm surprised to hear you say so. I'll tell you which I should prefer—Judge Buller. He is rather severe, it is true; but, at least, he would never blunder one's life away.'

He mentioned a remarkable anecdote, illustrative of a particular Providence, which may often be traced in tracking out murders. He had it, he said, from Lord Eldon :—' A man committed murder on his master, a farmer, in the stable of his dwelling between Cambridge and Huntingdon. He then rifled the house of some valuables, and, making his escape, went abroad, where he remained twenty years. A large reward was offered, and other active means were taken at the time for his apprehension; but the memory of the event and of everything connected with it, had long since passed away, when, at the end of twenty years, he returned, and on his landing went into an inn, in consequence of a violent shower. While he was warming himself at the fire, another man, also driven in by the rain, came and stood by him. They exchanged a few words, when the last-mentioned man went to the window of the room to watch the state of the weather. A hole in one of the panes was stopped by a piece of old newspaper, which caught his attention. It contained the identical advertisement, issued twenty years before, descriptive of the murderer's person, and offering the reward for his apprehension. It referred to some particular mark in his face, which struck the reader to correspond with what he happened to have observed on the face of the individual at the fire. This led him to read the description again attentively, and to go back and make his observations, when, in spite of the lapse of time, the general correspondence between the two was so apparent that he instantly went to the publican, strongly expressed his conviction, and urged him to take the stranger into custody. He did so, and the man was subsequently tried and hung for the murder.'*

He spoke with high commendation of Dr. Carey, the Baptist missionary, and traced him from his

---

* The gibbet on which he was executed was at Caxton, near the place where the crime was committed. The mound on which it stood, I understand from a friend, has been constantly kept up, and a new gibbet was some time since erected to commemorate the event. Caxton Gibbet is a well-known measuring place for distances.

lowly origin as a shoemaker to his then dignified position in the learned world. He was so unskillful at his trade that he could never make two shoes alike. He felt an ardent thirst for learning, and the piety of his mind directed his studies to Biblical literature. He was almost self-taught. After making a considerable progress in Greek and Latin he studied Hebrew; and being filled with missionary zeal turned his thoughts to India. The little Baptist society to which he belonged could at first raise only 13l. in support of the mission of which he was to be the head. From such small beginnings emerged a society which has since produced very striking and beneficial results to the cause of Christianity. The profits of his situation and of his literary labours, to the amount of 1,500l. per annum, he gave wholly to the mission. Upon this disinterested and noble appropriation of so large a sum Mr. Wilberforce poured forth an eloquent eulogium. Carey was no less active and zealous to promote the conversion of the heathen than to pursue the course of his learned labours.*

---

* In April 1801, Mr. Carey was appointed teacher of Bengalee in the College of Fort William, which had been established by Lord Wellesley in 1800, in Calcutta. The Rev. David Brown, Senior Chaplain, was Provost, and the Rev. Claudius Buchanan Vice-Provost.

In 1807, after the establishment of the College of Haleybury in England, the College of Fort William was reduced, and the offices of Provost, and of Vice-Provost, were abolished. But Mr. Carey, who

Lord St. Helen's told Mr. Wilberforce that during his embassy in Russia he travelled in the suite of the Empress Catherine to Kioff and other parts of her dominions. It was a sort of tour of inspection, and the authorities of many of the places through which she was to pass, in order to impose upon her, prepared along her line of road, quite in a theatrical style, various objects indicative of internal prosperity, which were to surprise and gratify her Imperial Majesty, and to which her attention was to be specially directed. Clusters of pretty snug huts, for instance, were constructed in various districts, with which the Empress was quite delighted; and, pointing them out to her attendants, exclaimed, 'See these people—what comfort they live in!' She had been flattered into thinking herself a great genius, and even wrote plays, or at least produced dramatic compositions of which she called herself the authoress. At the representation of one of these at Kioff, Lord St. Helen's was sitting with Prince Potemkin in his

---

had hitherto only ranked as a teacher, was now raised to a professorship.

In 1830 the Professorships of Fort William were abolished, and Dr. Carey was appointed an examiner, on a salary of 500 rs. per month.

Mr. Butterworth Bayley, a member of the Supreme Council, remarked 'that Carey's case was deserving of particular consideration, for in a service of thirty years no fault had once been found in him.'

Dr. Carey died June 9, 1834.

See Marshman's 'Life and Times of Carey.'

box, and was in the act of quizzing the piece when the Empress entered; and his Lordship, upon being interrogated by her, was forced by circumstances to allow that they were criticising the play. 'Well,' she exclaimed, 'if I don't succeed in pleasing others, I at least please myself.' Every morning at breakfast a new poem or sonnet was laid on the table, and said to be Her Majesty's composition.

Mr. Wilberforce sometimes expatiated with great animation on the superior and lofty qualities of Burke's mind; but observed that the House of Commons was not the scene in which they were most advantageously displayed. 'He was easily put out of temper, and there were those there who knew how to vex him and make him expose himself. When he was in good humour, and the subject on which he spoke suited him, and the House was in a listening mood, he was delightful. On such occasions there was a depth of thought and a fervour and glow of eloquence which defied competition. He had only to scratch the ground and flowers sprang up. Burke having been constantly opposed to Pitt up to the period of the French Revolution never became on a footing of intimacy with him; but he was always partial to Wilberforce, and used to dine with him in the course of every year to meet Windham. On such occasions he was most delightful and instructive, and fairly let himself out. He was so good-natured and fond of talking that, whatever his engagements were, he was apt to commence conversation and then to forget time. He would give up an hour of his own time, or consume one of yours, without reflection on consequences. Windham, though one of his greatest admirers, one day said of him: "When I have business to transact I avoid Burke as I would the plague." His papers,' added Mr. Wilberforce, ' used to be in greater confusion than my own.'

Speaking one day of the fine edge of Canning's humour, he observed : ' You see the joke sparkling in his eye before he gives it utterance. It appeared to me to furnish a sort of intellectual parallel to the natural fact that light travels quicker than sound— you behold the flash before you hear the report.' On another occasion he said there were moments when Pitt and Fox carried their auditors along with them with a power that appeared at the time almost irresistible ; ' but,' added he, ' so varied are Canning's qualifications—such his eloquence, wit, and humour —and so striking his figure and manner, that I really must account him on the whole as perfect an orator as I have ever known. Ah!' said he, ' Canning is now at the top of everything, and he is about to marry his daughter into one of our noble families— and Mrs. Canning will become a peeress—and—then —the bubble will burst.'

How soon these words were to be verified we little

thought! While they were uttering, the tomb was almost ready to open for this great and accomplished statesman. Often it has been my privilege to witness the exertion of those great oratorical powers which Mr. Wilberforce so highly eulogised, the transcendent merits of which are universally recognised.

Among contemporary statesmen, there was no one who entered the lists with Canning more successfully than Henry Brougham, now Lord Brougham and Vaux, whose extensive knowledge, great powers of debate, and indomitable courage, rendered him a formidable opponent. I cannot allude to this influential and now venerable nobleman, without offering a passing tribute to the zeal, ability, and untiring perseverance with which he ever supported the great cause of Abolition: and he mingled with it frequent demonstrations of personal esteem and regard to Mr. Wilberforce, which proved that he no less affectionately valued the man, than he honoured the statesman.

On another occasion Mr. Wilberforce said: 'Lord Grey's mode and style of speaking was, in his best days, peculiarly elegant; yet he was deficient in one of the leading qualities of a great orator. He is acute and logical, his language is pointed and his manner dignified; but he wants that fertility of imagination which is the soul of illustration. There is a somewhat repulsive harshness of manner occasionally about him, the consequence, in a great measure, of party spirit.' He then added: 'Grey never much liked me. He was aware of my intimacy with Pitt, and my aversion from party. One thing, however, there was about him which always pleased me; he appeared to be a man who acted upon principle. While many other persons would laugh together in private over principles and questions upon which they had attacked each other in public with apparent earnestness and sincerity, he never would.'

Mr. Wilberforce said to Lord Eldon, about the time of the second illness of George III. at the moment when the King again began to transact business: 'How is the King? I fear you and your friends have been in too great a hurry to bring him back to public life: surely he can hardly yet be fit for it.' 'Indeed, but *he is*,' said Lord Eldon; 'and I'll give you a proof how much he is himself. The first time I went to his closet after his recovery, to pay my respects, he called to me as I was retiring: "My lord, how is Lady Eldon?" "She is pretty well," I replied, "but I could hardly have flattered myself that she would attract your Majesty's notice or enquiries." "Indeed, but I am deeply indebted to her," replied the King, "since had it not been for Lady Eldon, your lordship would at this moment have been a country curate, instead of my Lord Chancellor."'

The King was perfectly aware that Lord Eldon, in consequence of his marriage, had given up his Fellowship at University College, and turned his thoughts from the church to the law.

The two following letters from Mr. Wilberforce reached me early in 1826 ; but I regret to say we did not return from London in time to receive either of these offered visits :—

'Beckenham: Feb. 28, 1826.

'My dear Friend,—The three months for which we came to this place have nearly passed away, and we are again meditating a visit to Bath. Shall you and Mrs. H. be at home in April or May? But let me not forget to ask another question. Should the tablet which I am to send you exhibit my own arms only, or Mrs. W.'s also? After receiving your answer there shall be no delay. Indeed, I consider your request as a mark of your regard, and as a means of recording our mutual esteem in your temple of friendship. One of my compeers, Acland, has just left me, after a summer's walk, so far as atmosphere can make it such—the birds also bear their part—the foliage only is wanting. How delightful must your place be to-day with all your evergreens.

'I am ever your sincere Friend,
'W. W.'

'Bath: May 17, 1826.

'My dear Friend,—Whenever we extend our drive into the country, in spite of the long drought, the sight of the whole vegetable world awakening from the sleep of winter and bursting into life and beauty suggests the thought that dear Mrs. H. and you must be beginning to think about returning to your own sweet place. We, also, are looking forward to an escape from city sights, and sounds, and smells. In about ten days more we meditate departing from this place, and fulfilling a long-made promise to spend a few days with a friend on the borders of Devonshire. We are also about to settle the time of a short visit to our dear friend at Barley Wood. But I should be very sorry to quit these parts without exploring your woods before the nightingales shall have left them. I therefore take up my pen to enquire whether you (always using the word in the dual number) are about to return home before it be much longer.

'I am ever sincerely and affectionately yours,

'W. WILBERFORCE.'

Early in 1827 we spent some pleasant days with Mr. and Mrs. Wilberforce, in Bath, and in the following autumn they were to have been our guests, but unforeseen circumstances on their part prevented it.

In the spring of that year I was in Paris, and he and his family spent part of the summer in Yorkshire, where his presence was warmly and enthusiastically welcomed by people of all classes. He was much touched by these tokens of affectionate respect and esteem, and he afterwards spoke with high gratification of the friendly intentions shown him by the venerable Earl Fitzwilliam during a visit which he and his family paid at his noble mansion, as also of the example set at Wentworth House by the daily habit of family prayer.

The following letter alludes to his Yorkshire tour:—

'Highwood Hill: Nov. 9, 1827.

' My dear Friend,—I suspect I am at least two letters in your debt, and that I have not even welcomed you home again to your own country. You probably know that not long after we last saw each other—having spent some little time with my excellent friend Babington, and visited our dear Bishop of Lichfield—I made up my mind to introduce Mrs. W. and such of my family as could accompany me to my few surviving friends in our great county. You will naturally hear with no small surprise that though for nearly thirty years member for Yorkshire I had scarcely ever gone into it except to York, on calls to public meetings, and Mrs. W. had only been within its verge at Rokeby, Morritt's beautiful seat, so honoured by Sir Walter's muse. The number of invitations which crowded in upon me compelled me to stay a shorter time than was desirable with each particular friend; but, on the whole, we have returned all highly gratified with our reception and with the opening prospects—and I can add, who knew the aspect of things forty years ago, with the highly improved state of the clergy, especially in the East and West Ridings. But more of this when we meet, when also I shall hope to hear more particulars of the general view you gave me of the improving state of the French Protestants. I am very sorry Madame Pelet was absent from Paris. All, I think, concur in speaking of her in the language which one should be led to expect by the exquisitely beautiful simplicity of her letters. I congratulate you on the opening of the South Wales College, at Lampeter, and on the share which Providence has allowed you in carrying forward that highly valuable undertaking. It is an honourable service. I hope dear Mrs. H. and you are well.

'I remain ever, my dear Friend,
    'Sincerely yours,
        'W. W.'

In the following year (1828) we again paid them a visit at Bath in the spring; and a visit to us was planned, but the state of his health towards the close of the year deprived us of this pleasure.

I paid him a pleasant flying visit at Highwood Hill, in company with Sir Thomas Acland, in the early part of this year—coming on him quite unexpectedly after ten o'clock at night—but as he had had his siesta he was in high spirits, and entertained us in his own delightful way till past midnight.

The two following letters will explain themselves:—

'Bath: Oct. 25, 1828.

'My dear Friend,—You will scarcely believe the list of arrivals correct in stating us to be at Bath so soon after your receipt of my last letter. And then, certainly, I had no idea of that rapid evolution. The fact, however, is that symptoms of impaired digestion led me to quit a friend's house where we were staying—and we arrived here on Saturday. I trust the water is doing me good, though I don't feel quite right yet. But for the state of ease and comfort I enjoy, as well as for innumerable other blessings, I may well be thankful. How little many years ago could it have been expected that I should enter my seventieth year, which I did on the 24th August last.

'It is not my system to pay visits after October. But the attraction to your abode is too strong to be resisted; and if you will be there, and can receive us, we hope to be with you in about three weeks. Perhaps Mrs. H. and you may pop over to Bath.

'Have you seen the last "Edinburgh Review" and the last but one? There are two articles on "History," and on Hallam's "Constitutional History of England," that are very extraordinary productions from young Macaulay.

'Hoping to see you ere long, and my eyes being far from equal to any beyond a very slight measure of penmanship, I will only now assure dear Mrs. H. and you of our cordial and affectionate regard. and of my being, with every good wish, my dear Friend,

'Ever sincerely yours,
'W. W.'

'Bath: Nov. 15, 1828.

'My dear Friend,—I have been for the last week indulging a growing growl against my own want of presence of mind, which (as sometimes happens to us in the case of the organs of hearing) I well remember I looked back upon as having been my condition a little after you and I had parted. I remembered that I had not seized, as I ought to have done, the declaration you threw out that you would bring Mrs. H. to spend a day with us, as we were unwillingly compelled to give up our visit to you. But surely you and I are not on such terms with each other as to render it necessary to be on the alert on such occasions—yet I grow alarmed at having come to the last day of the week without seeing or hearing from you. I therefore take up my pen to say that we shall be here to receive you and Mrs. H. any day next week; and I can truly add of all of us what I can most sincerely declare for myself, that there are scarcely any one (two rather) whom I should welcome with more cordiality. So I trust you will not disappoint us. My eyes being very indifferent, I travel over the paper *en galop*, and with kindest regards to your dear lady, am

'Sincerely and affectionately,
'W. W.'

When we parted at Bath, soon after the date of the last letter, it was settled between us that he should spend his seventieth birthday at Blaise Castle. In the course of the year we had paid him a short but delightful visit with Mr. Hart Davis, at Highwood Hill, and accompanied him in a long ramble to various beauties in the immediate vicinity. His conversation as usual was fraught with high interest, and his affection full and overflowing. Various letters were exchanged between us in the same strain of kind feeling as those already introduced, with the view of arranging his promised visit to us in August

1829. He made an effort to reach us, as had been settled, on his seventieth birthday, but did not arrive till the day after. I never saw him in more delightful spirits. The soul's calm sunshine and the heart-felt joy are his in a peculiar degree. No cares, no repinings, appear to have any place in his mind. Instead of recurring to the great part which he has acted in public life with self-complacency, or appearing desirous of fixing attention upon himself, his whole deportment bespeaks an entire self-renunciation—a deep and unfeigned humility—blended with a cordial kindness and a lively playfulness which are truly endearing. His mental powers continue vigorous as ever, and the flow of his conversation is unceasingly attractive. He has lately been reading Herschel's work on Astronomy, which has greatly fired his imagination: and he said some beautiful things suggested by it on the vibrations of light necessary for the production of each of the prismatic colours. His anecdotes and recollections of the past have also been, as usual, very amusing. Then his wit, his ready application of fine passages of poetry to the scenery of nature, the devout elevation of his religious reflections, and the touches of feeling and tenderness which break forth from him spontaneously have all combined to shed an enchantment on the hours of familiar intercourse with him.

The pleasure which I enjoyed in his society during

this visit was renewed in one which I shortly after paid him at Bath.

The two following are amongst the anecdotes above alluded to. When the Bank of England stopped payment, Mr. Wilberforce was at Bath. Pitt had requested his brother-in-law, Eliot, to apprise him of the fact, but he had previously heard of it by a letter from Admiral Bedford to Miss Patty More. He immediately went to call on Burke, who was then at Bath in a declining state of health. Windham and Dr. Lawrence were with him. Like everyone else, Burke at first took a gloomy view of the measure and of its probable consequences; but it is well known how much the issue dissipated these apprehensions.

Burke threw out many sensible remarks on the subject, and I was much struck by observing how entirely his two friends hung upon his words, as if every accent he uttered was oracular.

During the Rebellion in Ireland, in 1796, cruel means were often resorted to for the discovery of pikes concealed in the houses or about the premises of the peasantry. The subject was afterwards commented on in the House of Commons by Mr. Adair, who stated the alleged facts, and enquired if it was possible that such barbarities had been perpetrated? 'John Claudius Beresford, who rose to reply to the question, spoke,' said Mr. Wilberforce, 'with an honesty, force, and feeling, which I never can forget. "I fear," he said, "and I feel shame in making the acknowledgment—I fear it is too true. I defend it not; but it will, I trust, be permitted me to refer, as a palliation of these painful facts, to the state of my unhappy country when rebellion and discord had roused to the highest pitch all the stronger passions of our nature—anger, suspicion, terror." Lord Clare, in the House of Peers, when the same topic was introduced, took a very different course, and actually defended the alleged facts on the plea of necessity. "Well," he said, "and suppose it were so, surely" —and so he went on. I shall never forget Pitt,' said Mr. Wilberforce, 'on this occasion. He gave one of his high and indignant looks, and stalked out of the House of Peers.'

In one of our Sunday strolls he said: 'How painful it is to think that the feelings called forth by controversy upon religious topics should have been more fierce and unmeasured than upon almost any others. The odium theologicum has, in fact, been the bitterest of all hatreds. *Mentiris impudentissimè* is the elegant rejoinder of one of these combatants to his opponent.'

Touching on the purity and elevation of heart which should mark the Christian character, he said: 'The thing we are all of us—*you* and *I* and all—too much disposed to forget is, that Christians are to be

o

a peculiar people. We are too much inclined to appear to be what other persons are. One thing I often accuse myself of is the not seeking more diligently occasions of attempting to promote the spiritual improvement of others. It is a difficult point, but we should make it the subject of prayer. Such a thing, for instance, as our friend lending that good book the other day to Lord W——.'

A lady once said to him in my hearing: 'How happy Mr. —— and I should be to invite, while you are with us, any friends of yours whom you would like to meet.' He took her hand and said, smiling: 'That's very kind; but I so enjoy this friendly circle, and when I meet with *Tokay*, I don't wish to add water to it.'

Speaking of the value of truth he remarked: 'Truth is the moral gravitation of the universe.'

His allusions to classical authors, and his occasional quotations from them by way of illustration, were always most happy; and, when he aimed at giving peculiar force to a sentiment or a maxim, the point and terseness of his language could not be surpassed. As an instance of this, the topic of conversation one day being the misery to which Cowper, the poet, was exposed by his extreme sensibility at a public school, 'Yes,' he exclaimed, 'it was a sensitive plant grasped by a hand of iron.'

Cuvier's wonderful account of the antediluvian

mammoth, found entire within the last few years in Siberia, wedged into the ice, by which means the flesh and hair were in a state of preservation, was mentioned, when Mr. Wilberforce immediately said: 'It had only been hanging in Nature's larder for the last five thousand years.'

CHAPTER IX.

Mr. W. details some of the leading particulars of his Life.

THE foregoing pages will show, we trust, in some degree, how richly Mr. Wilberforce's memory was stored with anecdotes of men and things, and what an instructive and entertaining companion he was.

On various occasions he had communicated to me many interesting particulars of his early life; but in the course of some long walks we took together in this place he gave me a connected and detailed account of its whole course, up to the time of the formation of his religious character.

At the close of these conversations he said to me: 'Harford, I have entered thus particularly into a detail of these facts, because I wish to make you, and one or two more, depositaries of them.'

I lost no time in committing to paper what he said to me, which was as follows:—

'I lost my father when I was eight years old. My grandfather was still living. He was a person of much influence in Hull, and was very active and useful in aiding in its defence during the rebellion of 1745, when it was a place of much importance from being a depôt for military stores. My father was a younger son, and had been in the Baltic trade. After his death his business was managed for some time, for the benefit of the family, by Mr. Abel Smith, father of the present Lord Carrington. I was sent to a school at Putney, near London, when I was about twelve years old. It was a very indifferent school, where there was a little of geography, a little of classics, a little of everything taught. My father's elder brother, who had married a sister of Mr. John Thornton, had previously settled in London. He had a house in St. James's Place, and a villa at Wimbledon. I was much with them. My aunt was an admirer of Whitfield's preaching, and kept up a friendly connection with the early Methodists; and I often accompanied her and my uncle to church and to chapel. I was warmly attached to them both. They had no children, and I was to be their heir. Under these influences my mind was, even in these early days, much interested and impressed by the subject of religion. In what degree these impressions were genuine I can hardly determine, but at least I may venture to say that I was sincere. From the tenor of my letters, some of which are still in existence, my friends in Yorkshire became alarmed with the idea that I was in danger of becoming a Methodist. My mother was a very worthy woman, one of Archbishop Tillotson's school, who always went to church prayers on Wednesdays and Fridays, but at this time had no just conception of the spiritual nature and aim of Christianity. The apprehensions I have mentioned brought her up to town to fetch me away. I very deeply felt the parting from my uncle and aunt, whom I loved as if they had been my parents; indeed, I have scarcely ever felt more pain of mind than from this separation. My religious impressions continued for some time after my return to Hull, but no pains were spared by my friends to stifle them, by taking me a great deal into company and to places of amusement. I might almost venture to say that no pious parent ever laboured more to impress a beloved child with sentiments of religion than my friends did to give me a taste for the world and its diversions. In some respects it has perhaps conduced to my usefulness that I did not retain my early impressions. I might, probably, in that case have become decidedly Calvinistic in my opinions, and this would have given such a tincture to my views, that had I written my book on Practical Christianity under their influence, it would have differed materially from what has been the issue of my maturer judgment. Neither could I have come in for the county of York, for Christianity would have given me notions of humility which would have prevented me from aspiring to such a situation. I should not, therefore, have taken the means which ensured it.

'Hull was then one of the gayest places out of London. The theatre, balls, large supper and card parties, were the delight of the principal merchants and their families. They dined at two, and met at each other's houses for supper at six. Cards followed. I being the son of one of the principal merchants of the place was invited to almost every party, and was made a great deal of. At first all this was very distasteful to me, but by degrees I acquired a relish for it; and so the good seed was gradually smothered, and at length I became as thoughtless as any amongst them. My talent for singing made my company the more sought. In this idle way I spent my vacations, for my friends had sent me, on coming down to Yorkshire, to a very expensive school, at which three or four hundred pounds per annum were paid for each pupil. I went very early to the University, and became a member of St. John's College, Cambridge. On the very first night of my arrival I was introduced to as licentious a set of men as can well be conceived. They were in the habit of drinking hard, and their conversation was in perfect accordance with their principles. Though often mingling in their parties, I never relished their society—indeed, I was often horrorstruck at their conduct, and felt miserable. After the first year, I shook off my connection with this set and lived much with the Fellows of the College. What the society of Fellows of Colleges may now be I know not; but the set that I became intimate with at St. John's neither acted the part of Christians nor even of honest men to me. Their object seemed to be to make and keep me idle. If I occasionally appeared studious they would say to me, " Why in the world should a man of your fortune trouble himself with fagging?" I was a good classic, and acquitted myself well in some of our examinations; but I greatly neglected mathematical lectures, and was often told that being so quick of apprehension it was needless for me to toil. Card parties with the Fellows, and other amusements, consumed my time. All this while Gisborne,* who was at the same college, was fagging hard, and constantly attended lectures. Often did the tutors say of men like him, in my hearing, that they were mere saps, but that I did everything by real talent. The pernicious influences which thus acted upon me, and the almost unbounded use of money that I enjoyed, betrayed me into those habits which I can truly say I look back upon with unfeigned remorse.'

* The Rev. T. Gisborne, afterwards Rector of Yoxall, and the author of many valuable works.

I will pause for a moment in his own narrative to add that I gathered from what fell from him on this and other occasions that, although thus frivolous in his pursuits, and bent upon pleasure, he was preserved from licentious habits and infidel opinions. The Rev. Thomas Gisborne, above alluded to, in the course of a visit I paid him at Yoxall Lodge, described him to me as by far the most agreeable and popular man amongst the under-graduates of Cambridge. 'You might see him, he said, in the streets, encircled by a set of young men of talent, among whom he was *facile princeps*. He spent much of his time in visiting, and when he returned late in the evening to his rooms he would summon me to join him by the music of his poker and tongues—our chimney-pieces being back to back—or by the melodious challenge of his voice. When I did go in to him he was so winning and amusing that I often sat up half the night with him, much to the detriment of my attendance at lectures the next day.'

To proceed with the narrative:—' I was acquainted with Mr. Pitt at Cambridge, but we were not at that time intimate. He, indeed, lived in a higher set than myself; but in the winter after we had each left college I often met him in the gallery of the House of Commons, which we were both fond of attending. When I was about twenty I formed the scheme of representing Hull, and made an active canvass. A dissolution took place in 1780—about two days after I came of age—and I had nearly every vote in the town. Mr. Pitt came into Parliament in November, a few months after me, and our acquaintance quickly grew into intimacy. We lived much together in a club which was chiefly composed of young men who had quitted college about the same time, and who held the same political opinions.* We generally dined together when the House rose in good time, but still oftener supped together. Bankes was in the habit of entertaining us at his house in London, and I received them at Wimbledon, where I had a villa. We were both in possession of our fortunes. The rest of our set were chiefly younger sons, or eldest sons in expectancy.

'Pitt, who delighted in the country air, has accompanied me to Wimbledon for months together, after the House broke up. Even when the House rose late we usually drove over there, in order, at least, to sleep in the country. Hundreds of times I have roused him out of bed in the morning and conversed with him while he was dressing. In fact, I was at this time the depositary of his most confidential thoughts.

'In the autumn of the year 1783, after the breaking up of Lord Shelburne's Cabinet, in which Mr. Pitt

---

* Some of these particulars have been given with more detail in the account of Mr. Pitt.

had held office, I accompanied him and his brother-in-law, Mr. Eliot, on a tour to France, one of our principal objects being to acquire the habit of speaking French with fluency. At the dissolution which took place in 1784, in consequence of the resolution of the King to stifle Mr. Fox's India Bill and to support Mr. Pitt against the Coalition headed by Mr. Fox and Lord North, I went down to Yorkshire to attend the county meeting. I will tell you what I have hardly ever told anybody else, that I had then formed in my own breast the project of standing for the county of York, though to anyone else it would have appeared a mad scheme, for I hardly knew a single leading gentleman, except Mr. Mason. It was a very bold idea, but I was then very ambitious. The county meeting was attended by the Duke of Norfolk, Lord Carlisle, Lord George Cavendish, and many other men of rank and influence. They kept up in those days a great deal of state, and came on such occasions in their coaches and six. The meeting took place in the Castle Yard, and lasted from eleven or twelve till four, and everybody obtained a hearing. At its commencement the two parties opposed to Mr. Fox were divided, but I was the means of uniting them by a speech which I addressed to them with that end in view. I had considered the subject well, and may venture to tell you, as a friend, that I have scarcely ever spoken better than on that occasion. What I said made so deep an impression that the idea was immediately started of my becoming M.P. for Yorkshire; and yet it was very contrary to the aristocratic notions of the great families of the county to place the son of a Hull merchant in so high a situation.

'There was a great dinner after the public meeting, at which the Tories and the old Parliamentary Reformers coalesced against the Foxites; but, while the bottle was going round they disagreed, and matters were proceeding to extremities when I again addressed them at great length, and pointed out the inevitable consequences of division, and the paramount duty of uniting for the public good. Harmony was again restored, and both parties were so well pleased with this result, that one and another came to me and said, " We must have you for our member." Measures were accordingly taken to try the pulse of the county in my favour, and the result being favourable, I was put in nomination, and returned by a very large majority. This election, however, put me to a heavy expense, for there was a contest, though the opposition was not directed against me, but lay between the two other candidates. While it was going on I daily received letters from Mr. Pitt, who took a deep interest in the result. I afterwards six times represented the county of York, and even when I at last retired I might have been returned without opposition; but I was restrained from standing by many weighty considerations. I was reelected for Hull a few days before I came thus in for Yorkshire, and my friends of that place were somewhat displeased at my quitting them.

'I belonged, at this time, to no less than five clubs. I remember the first day I went to Brookes's, when I was just twenty, knowing scarcely anybody. Through mere shyness I played a little at the faro table, where George Selwyn, of fashionable notoriety, presided, keeping the bank. An acquaintance of mine, who saw me at play, and was aware of my inexperience, exclaimed, in a tone akin to pity, as though he viewed me as a victim dressed out for sacrifice, " What, Wilberforce! is that you? " George Selwyn seemed quite disconcerted by this interference, and, patting me on the back, said, " Don't interrupt him—he could not be better employed." Fox, Sheridan, General Fitzpatrick, the Duke of Norfolk, &c., were frequently there. Nothing could be more luxurious than the style of these clubs. The rooms were very handsome and splendidly furnished. Everybody met there on the easiest footing. You played at cards, or gambled, or talked, according to your fancy. There were two dinners provided daily; but altogether it was a profligate place. I belonged to White's also, which was much in the same style. The first day I went to Boodle's I won twenty-five guineas from the Duke of Norfolk. It was well for me that I never had the passion for play. After my views changed I took my name off all the clubs I belonged to, five in number, in one day.

'During the recess of Parliament, in October 1784, I accompanied my mother and sister, and two other female relatives, to Nice, by way of Lyons and Besançon, and Mr., afterwards Dean, Milner accompanied us. He and I were fellow-travellers in one carriage, and the rest of our party in another. Milner, though he held theoretically the same religious principles as in more advanced life, did not at this time act consistently with them; but was very much a man of the world in his manners, and was lively and dashing in his conversation. I myself had imbibed sceptical notions, and had hired a sitting at Lindsay's Unitarian Chapel. Milner, however, was a sincere believer, and when I let loose, as I sometimes did, my sceptical opinions, or treated with ridicule the principles of vital religion, he combated my objections, and would sometimes say, "Wilberforce, I don't pretend to be a match for you in this sort of running fire; but if you really wish to discuss these topics in a serious and argumentative manner, I shall be most happy to enter on them with you."'

Just before quitting Nice, in the winter of 1784 5, Mr. Wilberforce casually took up Doddridge's 'Rise and Progress of Religion,' and asked Milner what was its character. 'It is one of the best books ever written,' was his answer; 'let us take it with us and read it on our journey,' which they did carefully on their way back to England, and with thus much effect, that he determined at some future time to examine the Scriptures for himself, and see if things were stated there in the same manner.

'I returned to England for the meeting of Parliament in 1785, travelling post-haste, and on arriving at Dartford in the night was told if I chose to go on I might be sure of finding the House sitting. It sate late then on the Westminster scrutiny. I proceeded to town early the next morning, and drove to Pitt's, to stay a few days with him before I got into lodgings. Pitt was in bed. I went up to him and said, "Well, I am glad to find you are all going on well." "If you had come," replied Pitt, "a few days ago, perhaps you would have found me going on far from well." He alluded to a reverse which he had met with in the House, and its possible consequences.

'In the summer of the same year Milner and I rejoined my mother at Genoa. We went thither by way of Switzerland, and I have never since ceased to recur with peculiar delight to its enchanting scenery, especially to that of Interlaken, which is a vast garden of the loveliest fertility and beauty stretched out at the base of the giant Alps. In the course of this journey Milner and I resumed the subject of religion, and my former convictions were confirmed and deepened. We read parts of the New Testament together, when I pressed on him my various doubts, objections, and difficulties. The final result of our discussions was a settled conviction in my mind, not only of the truth of Christianity, but also of the Scriptural basis of the leading doctrines which I now hold.

'After my return home, serious reflections on the contrariety of my mode of life to these convictions would steal across my mind; and while in the full enjoyment of all that this world has to bestow, my conscience told me I was not leading a Christian life. I laughed, I sang, I was apparently gay and happy; but such thoughts as these would present themselves: "What madness is the course I am pursuing. I believe all the great truths of the Christian religion, but I am not acting as though I did. Should I die in this state I must go into a place of misery." Then I thought to myself, "And yet I may become religious. Has not God promised to bestow His Holy Spirit on them that ask Him?" Reflections like these led me to deep and earnest prayer. What followed was the natural consequence of enquiries pursued in this spirit. I was brought under deep convictions of sin. I looked back on my past life with unfeigned remorse. My anguish of soul for some months is indescribable, nor do I suppose it has often been exceeded. Almost the first person to whom I unfolded the state of my heart was Cowper's friend—good old* John Newton— whom I had often heard preach when I lived with my uncle and aunt. I had no other religious acquaintance. He entered most kindly and affectionately into my case, and told me he well remembered me, and had never since ceased to pray for me. Strange reports were now raised about me. I was said to be out of my mind and melancholy mad. These reports were conveyed to my mother and to my relations in Yorkshire, and for some time made them very uneasy; but on my going down to visit them, which I soon did, I took particular pains to be cheerful and pleasant and kind in the society of my friends. My natural disposition was irritable, and it had often cost them much pain. They were exceedingly struck by my altered deportment—they found me so much more kind and patient—so much more forbearing and considerate than formerly, that one of them remarked, if such were the effects of becoming "melancholy mad" it would be well if many of our acquaintance would take the infection.

'The year I was elected for Yorkshire I went to

---

* Rector of St. Mary Woolnoth, London.

*P

the races and attended the ball. I sang in public, and did everything to please my constituents. After I became religious I resolved never to be present again on these occasions. I should have been glad to have attended the grand juries, but could not without being involved in difficulties upon points of worldly conformity.

'Mr. Newton, in the interviews I had with him, advised me to avoid at present making many religious acquaintances, and to keep up my connection with Pitt, and to continue in Parliament. I wrote to Mr. Pitt, frankly communicating to him the great change that had taken place in my views, and the effects which this change would probably produce upon my public conduct. I told him, that although I should ever feel the greatest regard and affection for him, and had every reason to believe that I should in general be able to support his measures, I could no longer act as a party man. Pitt's reply was most kind. He assured me that nothing which I had communicated to him could ever alter our friendship, and that he hoped I would always act as I thought right. I had said to him that perhaps it would be as well that we should not when we met enter into any discussion upon the topics of my letter. To this he replied, "Why should we not discuss them?" He thought I was in low spirits, and proposed to come and spend the next day with me at

Wimbledon. He came, and we had a great deal of conversation. He was at that time inclined to be sceptical. The fact is he was so absorbed by politics that he had not allowed himself time for due reflection on religious subjects. At first he tried to reason me out of my convictions; but he found himself unable to question their justness, or the propriety of my resolutions, on the supposition that Christianity is true. I have entered thus particularly into a detail of these facts, because I wished to make you and one or two more a depositary of them.'

The following letter, a copy of which I received from a friend many years ago, addressed by Mr. Wilberforce to an individual in whose religious welfare he was deeply interested, confirms the correctness of much of the preceding narrative:—

'Oct. 25, 1809.

'My dear ——,—The letter which I have just received from you is far too interesting to me not to sit down and reply to it without delay—though I must write both more hastily and briefly than I should be glad to do. I am the more impatient to take up the pen, because I can with truth speak to you the language of encouragement. I should be sorry for your (mental) sufferings, were it not for the indication they afford me that you are in the right road. I myself have travelled it. You see and know that

I have long been as cheerful as anyone need be, and as active in discharging the duties of life; but I can truly assure you that when I was first awakened to a sense of the worth of the soul, and of the importance of Divine things, the distress I felt was deep and poignant indeed. I wondered, also, to find that my feelings so little corresponded with the convictions of my understanding. But in looking into the writings of many great and good men I found that their feelings and complaints had been exactly similar to my own. But, alas! my dear madam, must I confess my weakness? Even still I have too often occasion to lament that strange torpor and coldness which so ill accord with the mercies I have received, and the sense of obligation of which I am conscious. This leads me to prayer and to self-abasement—to penitential sorrow—to humble but earnest supplication for the promised aids of the Holy Spirit, for the sake of that Saviour who died upon the Cross to atone for our transgressions, in order to soften, to animate, to warm my dull heart. It is one of the promises of the New Covenant that the heart of stone shall be taken away, and the heart of flesh be substituted in its place; but this, as well as every other blessing, is only to be obtained through earnest, and to be preserved by continued, and recovered by renewed prayer. Meanwhile, my dear madam, it may well be a comfort to you that the promises of Scripture are made in the most express and positive terms to them that seek—to them that hunger and thirst after righteousness. Again and again is this blessed assurance repeated, doubtless because that Omniscient Being who sees the future as the present, foresaw that persons in this state of mind would be slow to take encouragement. It is therefore remarkable that the promise was twice made by our blessed Saviour's communications to St. John, recorded in the two concluding chapters of the Revelation : " I will give to him that is athirst of the water of life freely." By the water of life is meant the Holy Spirit's influence, as we know on our Saviour's authority. You will find some admirable remarks on the due measure of conviction for sin in a small book which I think one of the best that ever was written, entitled, "Witherspoon's Practical Essay on Regeneration." The change which is produced on the mind by a just view and right feeling of Divine things is in nothing more remarkable than in the different estimate which we are led to form of sin, of all sin, of all disobedience to, and neglect of, the authority and will of God. In truth, there is nothing which shows more plainly how little the religion of the world in general is the religion of the Bible than the slight and superficial sense commonly felt respecting all sin, except, as you truly say, of a few great and palpable instances of gross disobedience. And is there anything which so clearly proves the natural corruption of man, and how deeply it has affected both our judgments and our hearts? For is there a crime which is abhorred more than ingratitude? Yet we are apt to think little of the utmost forgetfulness, of the grossest ingratitude towards that God who created us—who all our lives long has supplied us with every comfort —who so pitied and loved the world as to send His only Son to die for us; and towards that Saviour also, who willingly left the glory and happiness of Heaven, and, after a life of hardship and contumely, endured at last a most painful and ignominious death, to rescue us from misery, and provide for us the means of obtaining everlasting happiness. Now, had any fellow-creature done for us a hundredth part of all this—had he prevented our suffering only a temporal death by consenting to die in our stead—had he only begged us to think of him with gratitude, and to conform our conduct in certain particulars to the rules he laid down, should we not deem ourselves monsters of ingratitude if we could live without scarcely ever thinking of this benefactor, or of the dying injunction He had left us? Yet of this ingratitude to our God and Saviour men in general, who call themselves Christians, and who really believe speculatively the great truths of Christianity, think little or not at all. A truly awakened conscience has a different view of things. It becomes sensible of its immense obligations, and of the poor return it has made for them, and it humbly devotes the future life to the service of that God, and to the honour of that Saviour, whose right to all our faculties and powers it now recognises, and whom it now feels it can never sufficiently love and honour and obey. These are in substance the words in which St. Paul describes the main principles of the Christian life. I quote from memory. "The love of Christ constraineth us," &c. Read also the remaining verses, 2 Cor. v.; and again, Rom. xiv. 7, &c.: "For none of us liveth to himself, and no man dieth to himself; for whether we live, we live unto the Lord," &c. Again, 1 Cor. vi. 20: "Ye are bought with a price: wherefore glorify God with your bodies, and your spirits, which are His." But I forget that my time is running away, and that I cannot leave altogether untouched the topics contained in the latter part of your letter. Here, also, my own case is applicable. But, before I speak from my personal experience, let me mention to you what was once really said to me by a woman—I might almost call her a lady—who, from being very dissipated and thoughtless, had become truly religious. On my asking her whether her friends and near connections relished her change of thinking and living she replied that she could not say they did; but then they said her temper was so much mended that it more than made up for it. But for myself. When, in 1786, I first became truly in earnest, and changed, I hope, the governing principles of my life from the desire of worldly estimation to that of pleasing God, it was reported throughout Yorkshire that I was melancholy mad, &c., and some of my friends were deeply wounded by these accounts. I happened to have occasion (the session of Parliament being over) to go into the country to spend a little time with my mother and sister, and a few of my oldest connections and friends, and when I first went to them they partook largely in this apprehension; but, after about a fortnight, I found that with one voice they declared to my poor mother, who had a horror of what the world calls Methodism, that if to go on as I did were to be a Methodist they could wish all whom they loved and valued to be such. I made it my special business to try to please them, so far as I could do it with a good conscience, and they were compelled to confess that my religion had rendered me more kind and obliging and self-denying than they had ever before known me. Depend on it this silent preaching—if I may so term it—this way of living down objections has a wondrous efficacy.

'It cannot be denied that much intercourse with the world has too often a sad effect in taking off the edge of our affections and the spirituality of our minds; but if we do not seek our happiness in it, but consider it rather as a scene of danger and trial, we shall doubtless be kept from suffering much from that degree of it which family life and duties may require. I should be very sorry to hear of your shutting yourself up in your chamber, and of your rendering your house dull to your son by preventing the innocent enjoyment of social intercourse. I remember an excellent friend and an experienced Christian was particularly earnest with me on this head—of conciliating the affections of my old friends, I mean such of them as were men of the world. I remember he pressed me and prevailed on me to go and stay two or three days, as usual, with Mr. Pitt, at Holwood. But when I say this, I am sure you will not so misunderstand me as to suppose that I do not mean that sufficient time should be reserved for devotional exercises, and that the maintenance of an habitual sense of the reality and importance of Divine and unseen things is not a most essential duty. It is often difficult in practice to draw any precise line between the allowable and excessive in recreations. I am glad you are not likely to be importuned to go into public; but there is a kind of regular dissipation prevalent, I believe, in other large towns as well as in London, which, from the time it consumes, as well as the useless irrational way in which that time is often spent, and the late hours that are kept, is extremely pernicious.

I must say for myself, however, that I have often found even the most rational society and the most innocent literary pursuits produce on my mind the effect of indisposing me to religious exercises—to prayer, meditation, &c. &c. Though I do not desist from them on this account, yet I think it requisite to guard against that effect with *jealous circumspection*. The perusal of good practical religious works is often conducive to this end. By the way, permit me to recommend and present to you a little book by the late excellent Mr. Newton, whom, after reverencing him as a parent when I was a child, I valued and loved as a friend for more than twenty years. He was the particular friend, also, of Cowper the poet; and the Letters, which are the work I mean, are more like Cowper's than any I know. Of course I do not mean to recommend every passage in them, but many of them are excellent, especially those to Lord Dartmouth, in the beginning of the first volume. I think you will agree with me in considering him as opening the heart surprisingly.[*]

'Before I conclude, let me mention to you one observation of great importance suggested to me by your saying you are reading and comparing the four Gospels. There prevails very generally a strange error, in which, however, I cannot conceive you to

---

[*] The Letters alluded to are published under the title of 'Cardiphonia.'

partake, that the Epistles are so difficult that it is better for Christians in general to confine themselves to the Gospels, and perhaps the Acts. I have for some years had an idea, only deferring the statement of it to the world from the want of time, of rectifying this dangerous error. I cannot explain my meaning in detail, but the general nature of it I will shortly lay before you. Doubtless the Gospels, especially that of St. John, which was written after the other three, contain all the doctrines of our religion; but these are given to us more in detail, and with a greater light of illustration, as well as followed out into their practical consequences more in the Epistles; for, in truth, it was not till after our Saviour's Ascension that the Apostles themselves were fully acquainted with them. Accordingly, we find that two of the disciples, in going to Emmaus after Christ's death, were utterly ignorant of the object of His death—the Atonement—and even conceived that it had refuted His claim to the character of the Redeemer of Israel. Our Saviour, in His last discourse to His disciples, which is recorded in those beautiful chapters, the 14th, 15th, 16th, and 17th of St. John, therefore told His disciples that "He had yet many things to say to them, but that they could not bear them now. Howbeit when He, the Spirit of truth, should come, He would guide them into all truth," &c. But one of the most remarkable instances of the difference I am speaking of, before and after Christ's death, is found in the same chapters, where our Saviour tells them, 1st, that hitherto they have asked nothing in His name; and 2ndly, that hereafter their prayers to God are to be offered up *in His name*. Accordingly, it has ever since been deemed the duty of Christians to offer up all their supplications and thanksgivings in the name of Jesus Christ. Now, can there be a more decisive difference than this? But it is the more extraordinary that the writings of St. Paul should be neglected by us Gentiles, because he was expressly commissioned by our Saviour Himself to be the Apostle, the instructor of the Gentiles; and therefore, though I put all the sacred writers on the same level, it might have been expected that the writings of this great Apostle would have been attended to by us with too exclusive a preference. But I must finish this unreasonably long letter, the former part of which was written amid innumerable interruptions. Let me, however, before I lay down the pen, beg you to excuse the egotisms of which I have been guilty, though I can scarcely excuse them myself, as well as the freedom with which I have addressed you; to which, however, I have been encouraged by the recollection of our old acquaintance, and still more by your own friendly frankness in opening to me the state of your heart. My dear madam, may the Almighty Himself grant you the best of all consolations; and I hope that after the depression of spirits produced but too naturally by the heavy loss you have sustained shall have passed away, you will have an internal source of never-failing comfort and happiness, even a measure of that peace which the world can neither give nor take away. In reading over to-day what I wrote yesterday, it has occurred to me that I scarcely said enough concerning those painful feelings which you laid before me. Besides the experience of the best men, whose interior has been unveiled to us, are not the Psalms full of similar passages, wherein the Psalmist laments the torpor of his spiritual affections, and calls on God to quicken him? There is a striking passage of this kind, I remember, in one of Barrow's sermons on the Love of God.

'Farewell, my dear Madam, and believe me, &c. &c.
'Your sincere Friend,
'W. WILBERFORCE.'

In this chapter we have epitomised, and, as nearly as we can recollect, in his own words, the particulars of the most important period of his life. Henceforth his object was to serve God in the Gospel of His Son, and to live to His glory.

In advanced life he recurred as follows to these events:—

'By degrees the promises and offers of the Gospel produced in me something of a settled peace of conscience. I devoted myself for whatever might be the term of my future life to the service of my God and Saviour; and, with many infirmities and deficiencies, through His help, I continue until this day.' *

* Life, vol. i. p. 112.

CHAPTER X.

1830 TO 1833.

Letters from Mr. W.—A Visit from him in November—He quits Highwood Hill in 1831—Visits to him at Bath—The Reform Crisis—Bristol Riots during a Visit from Mr. W.—In 1832 Domestic Affliction—Visit to Mr. Stephen—Visit from Mr. and Mrs. W. and Bishop Ryder—This was Mr. W.'s last Visit—Vigorous Letter from him in February 1833—His Speech at Maidstone in April—Letter from him at Bath in May—His Interview with Mr. J. J. Gurney—Leaves Bath for London in July—Illness and Death—His Funeral—And Last Honours paid to him.

THE beginning of 1830 brought me the following letter:—

'Highwood Hill: Jan. 23, 1830.

' My dear Friend,—So I find you have been fulfilling one part of Pope's description, by doing good by stealth; and if you will not be called on to verify the latter, by blushing to find it fame, you will earn a more gratifying recompense by the consciousness of co-operating at the same time in a work of charity that may continue to bless generations yet unborn.

' I suppose your neighbourhood participates in the

distress experienced by agriculturists even more, I believe, than by our commercial and manufacturing classes. My rents have of late been much lessened. Any personal feelings, however, are lost in those for our country in general. I really should like to know the sentiments of such a man as Mr. Hart Davis on the causes and probable consequences of these distresses. I have a high idea of his sagacity and observation. He has also been long enough on the stage to be able to look back upon the course of public affairs during the period of a very liberal circulation of paper, and after its being checked and reduced. I myself cannot but ascribe the distress of the farmers in a considerable degree to the sudden suppression of the country bankers' paper. And I must add, that of all the actions of the public men of our day I recollect none which appears to me so unreasonable, so almost utterly groundless, and perhaps so mischievous as Lord Liverpool's famous letter to the Bank of England. I seldom trouble you with my political speculations; indeed, to confess the truth, though politics were so long my trade they were never to my taste. How thankful have I often felt that it is so much easier to discover the path of duty than that of policy, and still more to decide how to secure our highest, our eternal interests, than our temporal. But my paper, and even more urgently my eyes, admonish me to lay down my pen. I hope

Mrs. H. is well. This house is so warmly situated, I have such sheltered walks, that I don't suffer from the wind whatever quarter it blows from. I overflow with blessings, and I thank God.

'Ever yours sincerely and affectionately,

'W. W.'

This letter was followed by the ensuing :—

'Highwood Hill : April 28, 1830.

'My dear Friend,—Let me transport myself, in this season of spring, to your rare exhibition of the sublime and beautiful. From a little spot of a dingle we have here, in which the violets (beautiful but not fragrant) and primroses are thickly sprinkled, I can form some idea of the delights of your paradise. It retains its Eve, too. When last there Mrs. W. and I walked on a Sunday by the stream, reading and musing over Montgomery's Hymns and the interesting essay prefixed to them, and I now endeavour to take that same stroll in idea at this season of the year, and to people your rocks and thickets (through my mind's eye) with all the wildflowers that they produce in such rich profusion. These objects should surely call forth in the coldest bosom warm aspirations of grateful and admiring praise. Do you happen to be acquainted with a volume of sermons lately published, and some of them excellent, by the Rev. Henry Raikes? But there is one of them, the views taken in which of the character of God would assuredly be contradicted by such a walk through your valley as I have been ideally enjoying. The writer in question too much restricts the beneficence of the Great Creator of the universe by confining it to *true believers*. Now, I cannot but feel most deeply Paley's exquisitely touching description of the infinitely varied delights of universal nature in all its varieties of animated being, where, following the bee from flower to flower, and seeing animals so happy that they know not how to give it expression, he breaks out into the apostrophe, "*After all it is a happy world.*" I shall not be so frequent an attendant at the anniversaries of charity societies this year as I formerly was. But I hope to attend the Church Missionary and the Bible Society meetings. When you come to town I trust you will pay us a visit. Farewell. Give my cordially affectionate regards to Mrs. H. and believe me,

'Ever yours,

'W. W.'

Though I was in town in the latter part of the spring of 1830, my impression is that Mr. Wilberforce was absent from Highwood Hill, and that we did not meet. In the autumn I received from him the following letter :—

'Isle of Wight: Oct. 4, 1830.

'My dear Friend,—To me it had appeared as well as to you *very, very* long since we had exchanged a word or a line; and it gave me no little pleasure to receive such a testimony of your friendly remembrance. Nor need I assure you, I am persuaded, that the account you give me of your dear relative's closing scene was deeply interesting to me. I have heard on good authority of many instances in which Christians truly deserving of the name, to whom death from a naturally nervous temperament had been through the whole of life an object of dread, were enabled at last to triumph over it, or were quietly released—in one, if not in two instances in sleep; and in another found in the attitude of prayer —from the posture of adoration on earth to the expression of it among the spirits of the just made perfect.

'Bath: Saturday.

'When I felt so comfortable in the Isle of Wight you will wonder at my not concluding my letter where I began it. Here we, however, are, conceiving myself likely to be benefited by a renewal of my annual potation of Bath water. I must say it is no very agreeable exchange to plunge into the smoke of Bath, from the pure air of the sea-shore and the Downs of the Isle of Wight, and the many-coloured tints, and venerable oaks, and picturesque hollies of the New Forest, in which we halted for a day or two with an old friend. It is one comfort that we are so much nearer to you. The little you threw out as to your movements was rather unpromising as to our meeting; but I trust "gravitation will not cease as you go by." I am sure attraction will not; and a double magnet, it has long been known, is most powerful. I hope your dear lady is well. Give us a few lines, my dear friend, without delay.

'Ever affectionately yours,
'W. W.'

Early in November we had the pleasure of welcoming our much-valued friends under our roof, and we had ourselves previously paid them a pleasant visit at Bath.

In the spring of 1831, Mr. and Mrs. Wilberforce quitted their residence at Highwood Hill; and amidst the preparations for their departure, recollecting that he had not replied to a question of some interest to myself which I had put to him some weeks before, I received the following note:—

'Highwood Hill: March 25, 1831.

'My dear Friend,—"Strike, but hear me," said the Spartan. I am almost glad not to be now within your reach. Yet sure I am, that having heard, you would own me guiltless. Never, even when M.P. for Yorkshire, was I more pressed by unavoidable claims on my pen than I have been for the last ten days. Correspondence about my chapel, my affairs, the settlement of my family previously to our quitting this sweet place for some years—perhaps for ever. I have been drawn from your Greek drama by the first rising of thoughts like these, having relation to events not quite so distant and interests not quite so touching. But really, my dear friend, I have read your Dissertation over several times, wishing to find something to correct or suggest, but in vain. It is true that I am not much acquainted with the subject, but the little that I do know enables me with confidence to praise it highly. In measuring the delay of my reply take into account the arrangement of books and papers, and letters and accounts—the accumulated stores of a whole life, and that a long one. I must break off. Farewell! With kindest remembrances to Mrs. H.

'I am ever affectionately yours,
'W. W.'

We twice visited Mr. and Mrs. Wilberforce at Bath, in May 1831, and I heard from him again in the month of September; and in October I received the following letter, in which he describes the symptoms of a malady, the renewed occurrence of which had for some time past caused his family and friends painful anxiety:—

'Bath: Oct. 7, 1831.

' My dear Friend,—Were I at liberty to obey my inclinations, I need not multiply assurances to convince you that I should at once thankfully say aye to your acceptable proposal, and hasten to act upon it. But I am bound in duty to make my health just now my main object. I have had of late several seizures, which, though we hope there is nothing serious in them, yet have so much of an alarming character that for hours I am utterly insensible. They enforce, therefore, the importance of attending to them without delay. The source of the evil appears to be the stomach; and for diseases of this class the Bath water is commonly found a peculiarly appropriate remedy. Mrs. W. and I, however, will indulge the hope of enjoying a week at Blaise Castle; and while I trust the exquisite features of your scenery will not have lost their charms, the still superior attractions of your social enjoyments will insure us pleasures of no ordinary character. This is an important week* for this country. May that Almighty Being who has so long favoured us still continue to bear with our provocations, and to hear

* The Reform Bill was being debated in the House of Lords.

the prayers of many among us who, I trust, do look up to Him with affiance and gratitude. We ought all to pray for our country, under existing circumstances, with double earnestness. The passage in Genesis, where ten righteous would have saved Sodom, and still more that in Jeremiah v. 1: "Run ye to and fro through the streets of Jerusalem, and see now, and know, and seek in the broad places thereof, if ye can find a man, if there be any that executeth judgment, that seeketh the truth: and I will pardon it"—are most cheering. *Sursum corda.*

'Believe me ever affectionately yours,
'W. W.'

The political crisis referred to by Mr. Wilberforce in the above letter occurred within a few days after its date, through the rejection of the Reform Bill by the House of Lords. During a visit which at this time I paid him at Bath he said he had long been of opinion that considerable changes and improvements in our system of representation, including an extensive modification of the borough system, were advisable and even necessary; but he objected to parts of Lord Grey's Bill, as being too sweeping in their changes.

Towards the end of October, Mr. and Mrs. W. paid us a visit of eight days, and were our guests during the occurrence of the disgraceful riots which reduced to ashes many important edifices of Bristol, and destroyed property to a great amount. These outrages broke out so suddenly that we knew nothing of them till they had reached their acme, which occurred on Sunday, October 30. It was not till after the close of morning service that we were made aware that serious disturbances had taken place in Bristol the evening before; but it was added that tranquillity had been restored. This, however, was far from being the case, for the removal of the military to a distance of some miles from Bristol, in consequence of the infatuated policy pursued by Colonel Brereton, the officer in command, gave full license to the ragamuffin mob of the preceding day, who, after various unchecked outrages, set fire to the Bishop's Palace and to other public edifices, as also to a large number of private houses and warehouses. The number of the mob swelled every moment, and they were allowed to break into the Bristol Gaol and Bridewell almost without resistance. We were first made aware of this alarming state of things as the evening set in by beholding the horizon reddened with the lurid glare of these devastating fires. The rioters were entirely put down at a very early hour on Monday morning, by the arrival of a body of cavalry, under the command of Major Beckwith, who charged at their head through all the principal streets with decisive vigour and success.* After the lapse of a few days I visited, in company with Mr. Wilberforce, the principal scene of the burnings and atrocities, and frightful was the spectacle it presented. The Mansion House and Custom House, two large buildings, were laid in ruins—here and there tall and singed walls menaced a sudden downfall—huge timbers, half-reduced to charcoal, stuck out of these walls, or lay heaped up among their ruins, which were still smoking. Many private houses, situated all around these public buildings, had shared the same fate and exhibited the same melancholy aspect. It was dangerous to move forward, for fear of the sudden crash of tottering walls; and it was appalling to stand still and to contemplate the surrounding scene—presenting, as it did, so fearful an illustration of the consequences attendant on the prostration of law and order, which had led to the surrender of an ancient and venerable city to the rage of an insane populace, many of whom perished in a drunken state among the ruins. Mr. Wilberforce was anxious to examine and localise every principal feature of the recent outrages; and this we were enabled, in some considerable degree,

* Before these troops came, Major (afterwards Sir Digby) Mackworth, acting on the spur of the moment, and heading a few followers, had already done much to stop the progress of plunder and devastation.

to effect by my knowledge of the city and of the events which had occurred.

These painful scenes interfered not a little with the tranquil enjoyment of social intercourse which I had promised myself with our venerable guest during this visit. He looked older and was more bending than when we last met. But his animation and cheerfulness were unabated, and his soul appeared to be vigorously exercising itself on heavenly truths and ripening for glory. He was, as usual, very entertaining and full of anecdote.

As events connected with the popular outrages above alluded to obliged me to be absent several mornings from our family worship, Mr. Wilberforce kindly consented to become our chaplain, an office he discharged, I was assured by several friends who at that time were staying with us, in the most impressive manner, and with a devout eloquence which was deeply felt by his auditors.

Early in the year 1832, Mr. and Mrs. Wilberforce were filled with painful anxiety about the health of their only surviving daughter. She had often been our guest, and the friendly regard we felt for her, no less than our sympathy in the feelings of her parents, led me, on hearing of her illness, to lose no time in writing to enquire after her.

Before the close of March the interesting object of

Mr. and Mrs. Wilberforce's parental solicitude was no more. They deeply felt the blow, but it was greatly attempered by the bright evidences which their daughter left behind of a blessed preparation for a better world. In the following letter Mr. Wilberforce briefly touches on it:—

'St. Boniface: March 26, 1832.

'My dear Friend,—I began a letter to you some time ago; but, could you know how I have been engrossed, you would not wonder that I had not been able to finish it. It was chiefly suggested by the interesting tract you sent me concerning the poor unhappy victims of justice at Bristol. I was sure dear Mrs. H. and you would sympathise with us on the afflicting scenes through which we have recently been passing. I will spare my half worn-out eyes by desiring our friend H. More to forward to you a long letter I have just now transmitted to her, containing an account of the last moments of her god-daughter. Blessed be God that I am able to send you such a narrative.

'Ever affectionately yours,
'W. W.'

I heard from him as follows under date August 10, 1832:—

'My dear Friend,—I would not wish you not to be angry with me for not having yet replied to your kind letter written so long ago. Yet I cannot wish you to be as angry at me as I am at myself. I could tell you a long story, and really a true one, of the various hindrances that have kept me from taking up my pen; but, after all, I do not pretend to justify myself. One thing, however, I will say, and that is, that the cause that has principally produced my silence has been that I felt I had so much to say if I should proceed to discuss the subject to which my mind naturally gravitated. I scarcely need say I allude to the learned volume you have sent me, though it is really one of its chief merits that, with all its scholarship, it is very interesting to a mere English reader—Mrs. W. for instance, who told me she could not lay it down again.

'I have been staying, for the last six weeks, with my dear friend and brother-in-law, Mr. Stephen, who a few years ago built upon the summit of one of the Buckinghamshire hills a comfortable though small house, which, for a pedestrian (and his health required constant walking), is one of the most delightful spots I ever knew. There are many adjoining Downs, the herbage of which is so short and fine as to be, where not cut up, a perfect bowling-green. These extend on the chalk from fifteen to twenty miles, and there are many beech-woods for shade on the sunny days, whenever one wishes a fit time and place for meditation. My friend has given to his place the name of "Healthy Hill;" and hitherto, even in this trying season, it has maintained its high character. We have had no appearance of cholera. Mrs. W. and I, having heard of its prevalence at Bristol, have written to enquire whether it has yet visited Bath. We pause before we decide on our movements. The good providence of our Heavenly Father is constantly watching over us wherever we are. Let us commit our ways to Him, and be at peace.

'Ever, my dear H., your sincere Friend,
'W. W.'

In the following October I received a letter from him at Bath, proposing to pay us a visit in company with Mrs. Wilberforce. We joyfully hailed his approach, and by a happy coincidence the Bishop of Lichfield and Mrs. and Miss Ryder were in the neighbourhood, and gave us the pleasure of their company during part of the time. This was the last occasion on which we ever had the delight and the honour of receiving our revered and loved friend under our roof, and it was one of the happiest of the many happy visits which he had successively paid us. He was in great vigour of mind, and in most animated spirits, full of affection and full of enjoyment. Sure I am that neither we nor our guests can ever forget the delightful conversations and intercourse which we enjoyed together, heightened and enlivened by Christian fellowship and affection.

During his stay the anniversary of our girls' school occurred, when he, together with several excellent clergymen who had breakfasted with us, accompanied Mrs. Harford and me to the schoolroom, and heard the children examined; after which, at our request, he addressed them in a most kind, affectionate, and impressive manner. It was interesting to observe how adroitly the veteran statesman and orator accommodated both his language and ideas to the capacities of young children; and in concluding he made them all laugh by saying, with reference to a set of reward bonnets and plum cakes just about to be distributed, ' Now, my dear children, don't eat the bonnets and put the cakes on your heads.'

Bishop Ryder's presence was no small addition to our enjoyment during this visit. He was a delightful companion as well as an eminent Christian. His piety was both elevated and practical. It was accompanied by a humility, gentleness, and cheerfulness which shed their benignant influence over his whole deportment, blended with that ease and grace of manners which belong to the best society. Happy himself, it was his delight to see others happy also, and his presence in the friendly circle, amongst his clergy or in general society, acted as a sunbeam. As a preacher he was truly impressive and faithful; and I never can forget the mingled energy and affection with which he brought home to the hearts and consciences of his hearers the great truths of the Gospel. During his residence at Wells, as Dean, he usually preached twice on Sundays, once at the cathedral, and again to crowded auditories in the large parish church, which was then destitute of any evening service, excepting when he thus officiated. He also exercised his ministry during the week at certain parishes within his Peculiar, as Dean, which were in a state of much spiritual destitution. It would be easy to dilate at great length upon his Christian zeal and benevolence; but suffice it to add that his memory will ever be fragrant in the affections of his friends and in the veneration of the church of Christ. Mr. and Mrs. Wilberforce on this occasion spent nine days with us, and the evening before their departure he was in such animated spirits, and conversed with so much energy, that after he had gone up to bed I said to a friend who was staying with us, 'I am truly glad you were present to-night. You have had a specimen of Mr. Wilberforce as I remember him in former years. Were you not delighted?' It was one o'clock before we separated, and he was then so ready to prolong the sitting as to say to me, 'Harford, will you read out to me one of those letters of Alexander Knox?' The letters in question were treatises rather than letters. I smiled and said,

'Are you aware of the hour, sir?' When I named it, he exclaimed, 'Oh! it cannot be.' I may here mention that his habit of taking a siesta after dinner restored him to his friends so refreshed that he was all alive and ready to sit up to a late hour, pursuing some interesting discussion, when they were ready to flag. During his stay we paid a long morning visit to his and our old and beloved friend, Hannah More, the particulars of which are given in Chapter XII. of this work, containing a brief sketch of that lady's life and character.

I cannot forbear inserting here a few passages from his 'Life,' illustrative of the undying energy with which he prosecuted to the last the great object to which he had so long devoted himself: On January 1, 1833, he addressed his old friend Macaulay, in glowing anticipation, as follows:—'I congratulate you, my dear old friend, on having entered on the year which I trust will be distinguished by your seeing at last the mortal stroke given to the accursed slave trade, and the emancipation of the West Indian slaves at length accomplished.' Following up this idea, we find him on April 12, in spite of his resolution never more to speak in public, induced to propose, at a meeting in the town of Maidstone, a petition against slavery, to which he affixed his own signature. 'It was an affecting sight,' say his sons, 'to see the old man, who had been so long the champion of this cause, come forth once more from his retirement, and with an unquenched spirit, though with a weakened voice and failing body, maintain for the last time the cause of truth and justice.'. . . ' There was now no question about immediate emancipation, but the principle of compensation was disputed, and on this his judgment and his voice were clear. He hailed, therefore, with joy the proposal to atone for the past offences of the country by the grant of twenty millions; and in this his last speech at once declared, " I say, and say honestly and fearlessly, that the same Being who commands us to love mercy, says also do justice; and, therefore, I have no objection to grant the colonists the relief that may be due to them for any real injuries that they may have sustained. But it must be after an impartial investigation of the merits of each case by a fair and competent tribunal. I have no objection either to make every possible sacrifice which may be necessary to secure the complete accomplishment of the object which we have in view; but let not the enquiry into this matter be made a plea for perpetuating wrongs for which no pecuniary offers can compensate." . . . " I trust," he concluded, " that we now approach the very end of our career ; " and as a gleam of sunshine broke into the hall, " the object," he exclaimed, with all his early fire, " is bright before us, the light of Heaven beams on it, and is an earnest of success." ' *

A short time after they left us I heard from him in his own vigorous style as follows:—

'East Farleigh, near Maidstone: Feb. 6, 1833.

' My dear Friend,—Though your letter of January 28th is not even now of a very old date, yet had any one told me immediately after my first perusal of it that it would wait a week for my reply, I should quite have resented the imputation—such was the warmth produced by the affectionate and friendly feelings you called forth by your account of our dear Hannah More, and the expression of your own sentiments and emotions in relation to myself. But the very day your letter reached me Mrs. W. and I went for near a week to our near neighbour and old friend, Lord Barham. It is a place which you will not be surprised to hear is always interesting to me, from my having inhabited it when occasionally it had been lent to me as a temporary residence, and still more frequently when, during a period of full forty years, the sweets of polished and liberal manners which I tasted in the company of its owners, were combined with the enjoyments resulting from a consciousness of mutual esteem and regard, grounded on a uniformity of principles in the most important par-

* Life, vol. v. p. 352-3.

ticulars, and gradually heightened into a real and affectionate friendship. I attended successively the funerals of two of these friends—Lady Middleton and Sir Charles, afterwards Lord Barham—and having known the present Lord, and indeed all that generation from their infancy (Lord B. himself in early youth lived with me sometimes for many months), a closeness of attachment has been produced which scarcely exists between any other family and myself. It is one of the many endearing circumstances of the place from which I write, that it is within a mile and a half of Barham Court, one of the very completest and most beautiful nobleman's, or rather gentleman's, places in England; though I am at once checked in my eulogium by the consciousness that it cannot, with all its beauties of wood and verdure, be at all compared with the romantic features of your own. I really hope that when you and Mrs. Harford come to London, or perhaps to *Ton*bridge (as is now the coxcomb mode of spelling it), you will allow us to profit from our vicinity to the great city, and have the pleasure of at least a short visit. I ought ere this to have stated that one reason, I believe the chief, which prevented my writing to you was my wishing previously to hear the sermon* of Robert Hall's which you had so

---

* On the glory of God in concealing.—R. Hall's Works, vol. vi p. 32.

highly eulogised. Mrs. W. began to read about half of it to me; but we were obliged to break off, meaning, as we still mean, to resume and finish it. It possesses all the dignity and pathetic sublimity of Hall, expressed in his own language with a fidelity quite extraordinary, considering that it is printed from notes. The particulars which Hall specifies as indicating the glory with which concealment clothes the Divine character appear to me just; yet it happened that my mind had recently been led into a different train of thought. I have been led to feel deeply how much we are called upon to admire and praise the kind condescension of the Almighty in acquainting us with His character and attributes in the degree which He has done. How embarrassing and anxiously distressing might it have been to our feelings if we had been left to judge of the qualities of the Supreme Being by the mere light of our unassisted reason. How gracious, then, is it in the Creator and Sustainer of the Universe to have quieted our apprehensions with such considerate kindness— to have soothed our misgivings and animated and assured our hopes by such varied means! It is only such majesty as that of God which can be seen without loss of dignity. Our human great men must, in order to retain their rank, prevent men's prying too closely into their qualifications and principles of action. But I must change my subject, for Mrs. W.

very justly impresses on me that in return for the kind interest you take in our children I ought not to omit their kind remembrances. I need not say how warmly Mrs. W. and I feel your kindness.

'Ever affectionately yours,
'W. WILBERFORCE.'

In the course of the five months which intervened between the date of the above letter and the termination of my revered friend's mortal career, I heard from him several times, and on May 21, 1833, he thus announced his arrival at Bath :—

'Bath: May 21, 1833.

'My dear Friend,—Here we are arrived once more to renew my potations. The place is very hot just now; but from the situation of our house I have shade more than one can commonly enjoy in summer anywhere. Life is so uncertain to a man who is not far from 74, that I should the rather be glad to recreate myself in your delightful habitation. Let us, therefore, form our plan with some knowledge of each other's—in other words, that I may form mine with due reference to your and dear Mrs. Harford's convenience, and the claims of your other friends. We arrived on Friday evening last. Of dear Hannah More we have heard as yet nothing: let me beg you to send me any intelligence you can. I hope all Mr. Hart Davis's family are well. Let me beg you to assure each of them of my friendly recollection and interest in their well-being. I should probably babble on, but am called away. So farewell, and believe me ever, my dear friend, sincerely and affectionately yours,

'W. WILBERFORCE.'

In a subsequent letter he proposed paying us a short visit in the last days of June, a proposal which we were most anxious to accept: but we were in London when this letter reached us, where illness and other unforeseen causes detained me so long as to frustrate our mutual wishes, and Mr. Wilberforce was obliged to leave Bath before we were in a condition to receive him. He did so on July 17, for Cadogan Place, Sloane Street, to consult Dr. Chambers. A little before he quitted Bath, Mr. J. J. Gurney, an excellent, highly respected member of the Society of Friends, called on him there, and has recorded some particulars of their interview, from which I copy the following passages:—' I found the veteran Christian upstairs, reclining on a sofa, his feet wrapped in flannel, and his countenance bespeaking increased age since I last saw him, as well as much delicacy. He received me with the warmest marks of affection, and seemed to be delighted by the unexpected arrival of an old friend.' Mr. G. then alluded to the brightness of his future prospect, when ' the illuminated expression of his furrowed countenance, with his clasped and uplifted hands, were indicative of profound devotion and holy joy. He told me that the text on which he was most prone to dwell, and from which he was deriving peculiar comfort, was— "Be careful for nothing; but in everything by prayer and supplication, with thanksgiving, let your requests be made known unto God; and the peace of God, which passeth all understanding, shall keep your hearts and minds through Christ Jesus" (Philippians iv. 6, 7). While his frail nature was shaking, and his mortal tabernacle seemed ready to be dissolved, this *peace of God* was his blessed and abundant portion. The mention of this text immediately called forth one of his bright ideas, and led to a display, as in days of old, of his peculiar versatility of mind. "How admirable," said he, "are the harmony and variety of St. Paul's smaller epistles! You might well have given an argument upon it in your little work on evidence. The Epistle to the Galatians contains a noble exhibition of doctrine— that to the Colossians is a union of doctrine and precept, showing their mutual connection and dependence—that to the Ephesians is seraphic—that to the Philippians is all love. With regard to myself," he added, "I have nothing whatever to urge but the poor publican's plea—God be merciful to me a sinner."' *

* Familiar Sketch by Joseph John Gurney.

'It was * (as his sons justly observe) altogether a striking combination of circumstances that he should have come to London at this time—to die. The Bill for the Abolition of Slavery was read for the second time in the House of Commons on the Friday night, and the last public information he received was, that his country was willing to redeem itself from the national disgrace at any sacrifice. "Thank God," said he, " that I should have lived to witness a day in which England is willing to give twenty millions sterling for the abolition of slavery." †

'Not less remarkable was it that London, which of late he had seldom visited, and where he proposed to remain but a day or two, should be the place of his departure. Yet had it been otherwise his funeral could hardly have presented the circumstances which made it the fit termination of such a life. The concurrence of two such incidents seemed providentially designed to fix public attention on his closing scene, that so the aged Christian might be marked out by the public voice as the man whom his country " delighted to honour." '

As it was not my privilege to witness the closing scenes of my revered friend's earthly existence, or

* Life, vol. v. p. 370.

† Lord Stanley (now Earl of Derby) in the course of his speech on this memorable vote, justly observed that Mr. Wilberforce on hearing of it might well exclaim, 'Lord, now lettest Thou Thy servant depart in peace.'

his sepulture, I here introduce some of the leading particulars of both from his 'Life' by his sons, in which they are depicted in a truly impressive manner, and which will form an interesting close to the foregoing pages, in which I have endeavoured to portray some of the characteristics of his social intercourse with a friend, who never thinks of him but with admiring and affectionate veneration:—

'On the evening of Friday, July 26, he expressed his gratitude to Almighty God that he was watched over in his last moments by his own children and by his own wife, all treating him with such uniform kindness and affection. On Saturday morning the fervency with which he offered up the family prayer was particularly noticed, and he seemed better. During that evening he became worse, and expired on Monday morning at 3 A.M. July 29, aged 73 years and 11 months.

'No sooner was his death made known than the following letter, originating with the Lord Chancellor (Brougham), was addressed to his youngest son, the only one of his four children who was with him at the time of his departure:—

"TO THE REV. H. W. WILBERFORCE.

"We, the undersigned members of both Houses of Parliament, being anxious upon public grounds to show our respect for the memory of the late

William Wilberforce, and being also satisfied that public honours can never be more fitly bestowed than upon such benefactors of mankind, earnestly request that he may be buried in Westminster Abbey; and that we, and others who may agree with us in these sentiments, may have permission to attend his funeral.

William Frederick	Grafton	C. J. London
Brougham, C.	W. Cantuar	Godolphin
Eldon	Wellington	Rosslyn
Lansdowne, P. C.	Ripon, P. S.	Calthorpe
Vassall Holland	Haddington	Bute
Westminster	Plunket	Denbigh
Clarendon	J. Lincoln	Ducie
Essex	E. Chichester	Caledon
Clifden	Bristol	Clanricarde
Wellesley	Gosford	Morley
Grey	Harrowby	Edward Hereford
Bexley	Albemarle	Dacre."
Sidmouth		

'In conveying this requisition, the Lord Chancellor declared himself authorised to add that nearly all the members of both Houses of Parliament would have joined, had the time allowed; and an application couched in the same terms was signed by almost one hundred members of all parties in the House of Commons.

'Mr. Wilberforce had chosen for the place of his interment, in accordance with a promise made to his brother-in-law, Mr. Stephen, a vault at Stoke Newington, where his sister and his daughter had been buried. A direction to this effect was given in his will: a circumstance, however, not actually ascertained till after the funeral. But his family had no hesitation in acceding to a request so gratifying to their feelings. Still they thought it fitting to avoid all such parade as was inconsistent with the situation of a private gentleman. It was his characteristic distinction that, without quitting the rank in which Providence had placed him, he had cast on it a lustre peculiarly his own. Nothing, therefore, could be more appropriate than that the bishops of the Church, the princes of the blood, the great warrior of the age, the King's chief servants, and the highest legal functionaries—whatever England had most renowned for talent and greatness—should assemble, as they did, around his unpretending bier. His simple name was its noblest decoration.

'When his funeral reached Westminster Abbey, on Saturday, August 5, the procession was joined by the members then attending the two Houses of Parliament. Public business was suspended—the Speaker of the House of Commons, the Lord Chancellor, one prince of the blood, with others of the highest rank, took their place as pall-bearers beside the bier. It was followed by his sons, his relations, and immediate friends. The Prebendary then in residence, one of his few surviving college friends, met it at the Minster gate with the Church's funeral office; and, whilst the vaulted roof gave back the anthem, his body was laid in the north transept, close to the tombs of Pitt, Fox, and Canning.

'It was remarked by one of the prelates who took part in this striking scene that, considering how long he had retired from active life, and that his intellectual superiority could be known only by tradition to the generation which thus celebrated his obsequies, there was a sort of testimony to the moral sublimity of his Christian character in this unequalled mark of public approbation; for, while a public funeral had been matter of customary compliment to those who died in official situations, this voluntary tribute of individual respect from the mass of the great legislative bodies of the land was an unprecedented honour. It was one, moreover, to which the general voice responded. The crowd of equipages which followed his funeral procession was unusually great. The Abbey was thronged with the most respectable persons. "You will like to know," writes a friend, "that, as I came towards it down the Strand, every third person I met going about their ordinary business was in mourning." A subscription was immediately opened among his friends in London: it was agreed to place his statue in Westminster Abbey, and, as a yet more appropriate memorial, that some charitable endowment should perpetuate his name.* Public meetings were held at York and Hull on the occasion; and in the former place a County Asylum for the Blind has since been founded in honour of him, while his townsmen at Hull have raised a column to his memory.

'It would be vain to mention all the marks of respect which were paid to him by the public societies in which he had borne part; nor were there wanting other more private, but not less affecting, tokens of regard. " Great part of our coloured population, who form here an important body," writes a dignified clergyman from the West Indies, " went into mourning at the news of his death." The same honour was paid him by this class of persons at New York, where also an eulogium (since printed) was pronounced upon him by a person publicly selected for the task; and their brethren throughout the United States were called upon to pay the marks of external respect to the memory of their benefactor. For departed kings there are appointed honours, and the wealthy have their gorgeous obsequies: it was his nobler portion to clothe a people with spontaneous mourning, and to go down to the grave amid the benedictions of the poor.

'It is impossible to conclude this history without

* What has been finally done in pursuance of the resolutions of the above-mentioned meeting will be seen by referring to a postscript, page 323, at the end of this volume.

observing the striking testimony which it bears to that inspired dictate: " Godliness has the promise of the life that now is, as well as of that which is to come." If ever any man drew a prosperous lot in this life, he did so who has been here described. Yet his Christian faith was from first to last his talisman of happiness. Without it, the buoyancy of his youthful spirits led to a frivolous waste of life not more culpable than unsatisfying. With it came lofty conceptions—an energy which triumphed over sickness and languor, the coldness of friends, and the violence of enemies—a calmness not to be provoked —a perseverance which repulse could not baffle. To these virtues was owing the happiness of his active days. Through the power of the same sustaining principle his affection towards his fellow-creatures was not dulled by the intercourse of life, nor his sweetness of temper impaired by the irritability of age. A firm trust in God, an undeviating submission to His will, an overflowing thankfulness: these maintained in him to the last that cheerfulness which this world could neither give nor take away. They poured even upon his earthly pilgrimage the anticipated radiance of that brighter region to which he has now doubtless been admitted. For " the path of the just is like the shining light, which shineth more and more unto the perfect day."'

CHAPTER XI.

Incidental Remarks on his Character and Domestic Habits—
Portrait by Lawrence, and ditto by Richmond—His Statue in
Westminster Abbey—Letter from Provost of Oriel.

MR. WILBERFORCE has been described in the preceding pages as one of the kindest and most social of men. It will have been seen how much he delighted in calling forth the mental resources of his friends, as well as in imparting to them the rich stores of his own mind upon subjects of mutual interest, and especially upon the highest of all topics.

Our social hours were enlivened by his wit, and by the innocent playfulness of his brilliant fancy. His whole manner and expression of countenance were the index of the sunshine of a mind at peace with God, and cherishing no thoughts but of kindness and love to his fellow-creatures. No one could be in his company without hearing noble sentiments, or allusions to schemes of benevolence, or expressions of sympathy and kindness propounded in the silvery tones of his voice, in the most natural and easy manner. He made a point of always being present at our family prayers. Having often, when staying with him, admired his devout and sublimated spirit in prayer on similar occasions, I sometimes tried to persuade him to act as our chaplain; but his constant rule was never to take this part in a friend's house, except in a case of sickness, or the absence of the head of the family.

It is to be regretted that no really characteristic portrait of Mr. Wilberforce was executed while he was in the prime of life; but he yielded to the wishes of his valued friend, Sir Robert Inglis, by sitting, when he was in his sixty-ninth year, to Sir Thomas Lawrence, and although at the death of that able artist the picture was left in a very incomplete state, yet the head was wrought up to a high degree of finish, and is an instance of the success with which Lawrence often caught the finest expression of his subject without any sacrifice of its identity. The intellectual power and the winning sweetness of the veteran statesman and philanthropist are happily blended in this portrait.

Sir Robert judged rightly in not allowing a single stroke to be added by another hand. It hung in his dining-room for more than twenty-five years, skilfully lighted by a shaded side lamp, and seemed to smile benignantly on the many friends who sat round that hospitable board.

This picture, thus lighted, bespoke the feelings of

reverence and affection with which its owner ever regarded the character and name of Wilberforce; while he, on his part, highly honoured and valued the eminent virtues, both public and private, of Inglis, and greatly loved, in common with all his friends, the bright social qualities and that perpetual flow of kindness and of information which won for him the regard of all parties, even of those who most essentially differed from him in the political views and principles to which, as a public man, he inflexibly adhered.

Sir Robert bequeathed the picture to the nation, and it now hangs in the National Portrait Gallery.

A full-length likeness of Mr. Wilberforce, also painted for Sir Robert Inglis, by George Richmond, R.A., in water-colours, represents him with characteristic fidelity. It has been widely diffused throughout the country by the medium of an engraving by Finden.

It is to be regretted that the statue raised to his memory in Westminster Abbey has not been equally successful. It gives undue prominence to the singularity of his figure in later years, and it fails to give expression to that amenity of feeling, and that bright intelligence which often played over his countenance and charmed his friends, even in the absence of all beauty of feature.

After morning prayers he often paced the terrace

of the house, or strolled into the flower-garden, and whoever was his companion was sure to see him full of delight at the various beauties or wonders of nature. Sometimes he was attracted by its grander features; at others, by the scent or the pencilling of a flower. Of flowers he was peculiarly fond. He delighted to gaze upon their colours, and to investigate their structure; and most of his favourite pocket authors were thickly set with them in a dried state. It was often hard to persuade him to quit the garden for the breakfast-table, and when he made his appearance it was generally with a flower in his hand. Once there, he was sure to be the life of the party. It was a meal in which he took particular pleasure. He was then also peculiarly ready for conversation and discussion, and we frequently forgot time when thus engaged. He used to say it was one of the distinctions between us and animals, that the latter sat munching their food by themselves, but that men have the faculty of exercising their mental powers while they satisfy the requirements of nature.

He was fond of receiving visitors at his own house at breakfast, or of meeting them at the houses of his friends. From the breakfast-table he retired to his dressing-room, and was actively engaged with his secretary in answering letters, or in listening to reading. Letters of importance, or to his more particular friends, he wrote, when his eyes permitted, with his own hand, but others were chiefly dictated to his secretary. About two or three o'clock in the day he generally came down stairs, and was ready for a drive or a walk; but a long walk was his particular delight. He possessed, as a companion, as we have already said, an unceasing fund of delightful conversation. A life spent, as his had been, amongst many of the greatest men of his time—himself one of the most distinguished—and possessed of social qualities of the highest order, how could the conversation of such a man be otherwise than deeply interesting? Then there was such simplicity and purity of heart—such touches of tenderness and of exalted piety — that every previous sentiment of affection and veneration towards him was deepened by being admitted to his familiar intercourse. He liked an early dinner, and throughout the whole course of it his mind was actively at work. At his own house I have often been amused to observe him carrying on a discussion while in the act of carving a large joint—often addressed by others, and often interlacing his subject with exclamations at the bluntness of the knife, or paying little attentions to his guests, and then taking up the broken thread of his subject and pursuing it amidst the same sort of interruptions. For the last eight or ten years of his life he was in the habit of retiring soon after the ladies left the room to take a siesta; but he reappeared at the tea-table, when his entry enlivened every countenance. He was fond of sitting up late with a friend, either for reading or talking. The habits of the House of Commons had made late hours so familiar to him that he often appeared reluctant to retire.

From youth to age Mr. Wilberforce assiduously devoted himself to his parliamentary duties, but throughout the prime of his life the great object for which he toiled and laboured was, as has been already shown, the abolition of the slave trade. Conjoined with this, however, he was the active and energetic supporter of a large part of the most important measures promotive of public improvement and benevolence, both in and out of Parliament, whilst a constant stream of private charity flowed forth from him in quarters innumerable. I never beheld in any person a more positive rejection, I might say, spurning, of anything like flattery, or a more unaffected renunciation even of that sincere homage which was often proffered by those who approached him for the first time. The sacred influence which had descended upon him from above early in life, when God called him by His grace, accompanied him through every stage of his earthly existence.

As a little instance of his consideration for the feelings of others, I had been folding up several letters in covers for him ready to be franked. On handing one of them to him, in a cover which had served the same purpose before (for he often used covers in good condition a second time), observing it to be ticketed with the name of an individual in humble circumstances, he said to me, 'Pray be good enough to envelope this letter in a clean half-sheet. If it was to anyone of the higher class it would do perfectly well; but this poor man may, perhaps, construe it into an intended slight.'

It has been observed that the possession of power is one of the severest tests of the human character. Men who have appeared in private life amiable and unassuming have often, when placed by the course of events in circumstances of high influence and station, become arrogant and presuming; and even good men, under such circumstances, frequently hurt the feelings of their friends by somewhat of the bustle of self-importance or the abruptness of decision. Not seldom have I been struck, in the course of my long acquaintance with him, with his utter superiority to all such littleness. Whether I saw him in the privacy of his own house, or in the lobby of the House of Commons—whether he was getting into his carriage for an ordinary drive, or on such an occasion as a visit to the Emperor Alexander—he invariably appeared in the same Christian frame of mind—intent upon the day's business; but lending a ready ear to any present appeal of humanity, or to any claim of friendship—animated by affectionate, kind, and cordial feelings, and desirous to lose no opportunity of receiving or of doing good. I have already spoken of his constant self-recollection, even amidst the busiest and most exciting scenes. Hence arose a no less constant self-command. Frequently have I seen him placed by forgetfulness or inattention in his servants in circumstances peculiarly harassing, at the very moment that he was pressed upon by important business, or by the urgent claim of an immediate appointment; and yet I do not ever recollect to have witnessed anything of unkindness or petulance in his manner of addressing or reproving them. Yet he has told me that he was naturally irritable to a degree which was very trying to his nearest relations. One of the very first victories which he achieved over himself, he added, after he became a religious man, was the entire correction of this bad habit. He became, and he continued through life, a model of the very opposite qualification.

In advanced age the fine powers of his voice naturally declined; but they were still such in reading, in speaking, and in conversation, even to the last, as greatly to heighten the effect of anything which fell from his lips. At public meetings of our great charitable institutions—for instance, the British and Foreign Bible Society, the Church Missionary, and others—his powers of eloquence often shone forth with a lustre which excited enthusiastic plaudits. To the last day of my life, I shall never forget some particular instances of this description.

The following is an instance of happy illustration. At a meeting of the London Society for Promoting Christianity among the Jews, a speaker objected that its success was insignificant in the great work it had undertaken. Mr. Wilberforce replied, 'It may be so in appearance; but were we to argue from apparent insignificance, I might point to the little infant in his wicker basket on the borders of the Nile. Here was apparent insignificance, yet this little infant became the great deliverer of the Jewish nation, and their Divinely-appointed lawgiver.'

Mr. Wilberforce was a sound Protestant, and an attached member of the Church of England; but when he had to unite with Dissenters in opposition to the slave trade, or in furthering societies or objects promotive of public utility, or practical benevolence, he did so with a cordiality and kindness of heart which proved how superior he was to an exclusive or sectarian spirit. Dissenters met him in a similar spirit; and I have often been struck when he entered a meeting, where many such were present among the auditors, by the affectionate enthusiasm with which they were sure to greet him. His politics were Conservative, yet Liberal, and whatever was the subject, whether political or religious, which he spoke upon, his moral courage and independence of spirit never failed him. These were qualities which made it impossible for him to be a party man, and it was precisely because he was not such that there was scarcely any man who exercised so extensive an influence for good as himself throughout the length and breadth of the land.

Of Mr. Wilberforce's parliamentary eloquence, very strong testimonies will be found in his 'Life.'* Amongst them, perhaps none conveys a more striking impression of it than that of his friend, Mr. Morritt, of Rokeby, who tells us that, when speaking on an important subject, he usually mingled with the staple of his general argument passages of such brilliancy as any speaker might have envied.

When an encounter was forced upon him, he proved himself to be a complete master of the science of defence; and the resistless power with which he once turned a personal attack of one of the ablest speakers of the House on his assailant is not yet forgotten in the annals of parliamentary retort.

Mr. Pitt, it is well known, said of him, that in point of natural eloquence he had never known his equal. The effect of his oratory was greatly heightened by the tones of his voice, which in his youth

* Life, vol. v. p. 241, &c.

procured for him the title of the nightingale of the House of Commons.

The following letter, recently addressed to the author, by the Rev. Dr. Hawkins, Provost of Oriel, gives so pleasing a picture of the impression produced on him by casually meeting Mr. Wilberforce in 1815, that, with his permission, it is subjoined to the preceding pages:—

'Oriel College: Nov. 2, 1864.

'My dear Mr. Harford,—In the spring of 1815, having got into a stage coach in town to return to Oxford, I found the coach stop at Kensington Gore, and a gentleman in black came into the coach with a "Quarterly Review" in his hand, an old-fashioned servant in livery getting also upon the roof. After a short time we got into conversation; he laid the "Review" aside, and a most agreeable journey I had with him from Kensington to Nuneham, where he stopped to visit a son who was then a pupil of the late Mr. E. G. Marsh, at Nuneham. No one else came into the coach during the whole journey; but I was exceedingly puzzled to guess who, or what, my agreeable companion was. At first I thought from his dress and his conversation that he was a clergyman; but he showed such an intimate acquaintance with the political world, and the great world, that I gave up that idea, and was quite at fault.

'I think he spoke of Pitt, I know he spoke about

the slave trade; for I had spent the preceding winter at Paris, and had seen a good deal of French society, both of those who had returned to France in 1801 or 1814, and of the revolutionary party (it was on the return of Buonaparte from Elba, indeed the day before he re-entered Paris, that I quitted it), and my companion was anxious to know what the French really thought of abolition. One thing which particularly struck me in his conversation was the singular elegance and accuracy of his *English*, whilst it was without the slightest tincture of pedantry. Well, when he left the coach at Nuneham, he desired me to call upon him at Kensington Gore, upon which I begged to know his name, and he said, "Wilberforce." " *The* Mr. Wilberforce?" I asked, and he said, " Yes." So I lost no time in calling upon him, and was several times a guest at his breakfast and dinner table at Kensington Gore . and at Bath; and at Highwood Hill in 1828 I spent some days with him. The last time I saw him was, I think, in 1831, when he spent a few days with us here. I have many delightful recollections of him, and often wish I had made memoranda, as you have, of what fell from him, as well as from some other remarkable men with whom I have had the pleasure of being acquainted.

'I dare say I have told you this before; but I mention it now in order to show you with what interest I must have read your "Recollections of

Wilberforce;" so much so that I must needs thank you for the pleasure I have derived from your book: I ought to say *profit* also; for it is at least my own fault if I have derived no profit from reading the story of so good a Christian as your admirable friend.

'You have drawn forth a long letter from me, and not on *business*; and I am so bad an economist of time, and yet have so many things to do, that I seldom write any other than business letters. Pray let it draw forth a few lines from you to tell us how you both are.

'With our united kind regards,
    'Ever yours most sincerely,
      'EDWD. HAWKINS.

'J. S. Harford, Esq.'

CHAPTER XII.

Brief Recollections of Mrs. Hannah More.

ON more than one occasion we have referred in the preceding pages to visits paid by Mr. Wilberforce to his old and much-valued friend, Mrs. Hannah More. I cannot quit this subject without giving a brief sketch of that eminent lady, and of her sisters, such as they were on my first acquaintance with them. I had, in fact, the most intimate means of judging of their characters and pursuits. They were five in number, and in their style of dress and manners belonged to a state of society which has long since passed away. Hannah More was the centre around whom they moved, regarding her with a degree of deference and affection which could scarcely be surpassed. My acquaintance with them commenced in 1809, by means of a letter of introduction from a friend whom they very highly esteemed. I reached their beautiful residence, at Barley Wood, early in the morning on a fine day, in company with the Rev. Walter Trevelyan, vicar of Henbury, second son of the late Sir John Trevelyan, Bart., who shared with me in the wish to see the celebrated authoress of so many works of genius and piety. We found ourselves ensconced in their breakfast room before any of the sisters had come down stairs, and, after waiting a little while, the door opened, and a lady entered, of animated manner and appearance, whose first words were, holding up both her hands, '*I am not Hannah More.*' This was Patty More, the youngest of the sisterhood, whose bright blue eyes and benevolent expression greatly pleased us both. She it was who, together with Hannah, was the life and soul of the great educational movement carried on under their auspices throughout Cheddar, Nailsea, and other neighbouring parts of Somersetshire—a movement which the pen of Patty has so admirably described in a small book, entitled 'Mendip Annals,' &c. They were first impelled to this good work by the exhortations and the purse of Mr. Wilberforce, who had visited Cheddar from their house, and had brought away with him a deep impression of its romantic beauties, and of the extreme ignorance of its population. The door opened again, and as another lady sailed in, Patty exclaimed, '*That is not Hannah More.*' This was Sally, the wit of the family, and a most kind-hearted and original character—of whom Hannah More used to say, that if her discretion had been equal to her wit, her good sayings would have been everywhere quoted. A third time the door opened, and a lady entered whose brilliant eyes lighted up a pale and sensible countenance which immediately bespoke our interest. Patty instantly exclaimed, '*That is Hannah More.*' Her reception of us was most frank and cordial, mingled with a kindness which made us feel quite at our ease with her. There was no effort to shine; but again and again pointed and bright things fell from her lips in the most easy and natural manner. We now took our seats at the breakfast-table, and were quickly joined by the two remaining members of the sisterhood, one of whom—Mrs. More, as she was called— the eldest of the five, was of somewhat stately appearance, and a little reserved in her address and deportment. She wore a high cap, surmounting a large ruff of powdered hair. Betty More, the remaining sister, was good nature and hospitality itself, and was never satisfied without pressing upon her guests all the good things on the table. All the details of the household were under her special management; and plenty and good cheer followed in her train. The trellis around the windows of Barley Wood was a temple of Flora, under the care of Sally, whose adjoining flower-beds were profuse of brilliant colours and fragrant odours. The love of gardening, however, occupied Sally's mind only in subordination to higher pursuits. She wrote some of the tracts in the 'Cheap Repository,' and was the life and soul of her Female Benefit Society, which did great good in Wrington and its neighbourhood. Its anniversary took place at Barley Wood, where the members were kindly welcomed and regaled; and on these occasions Sally would mount a rural rostrum, and charm her auditors by an address, in which a great deal of good advice was set off by much of that original humour peculiarly her own, which always sparkled in her merry eyes.

Conversation was kept up in the most sprightly manner; and though it was difficult to answer all the questions and statements which pressed upon us in the same breath, from one and another of the circle, it was quite clear that each of them was intent on exalting our ideas of Hannah More, and of bespeaking our attention to every sentence that she uttered. She, on the other hand, was no less anxious that her eldest sister, who was rather deaf, should lose nothing amusing or interesting which enlivened our meeting. They insisted on our spending the day with them, which we gladly did. Some part of it was passed by my companion and myself in visiting Brockley Combe, at that time one of the most romantic glens in England, though its rocky acclivities have since been denuded of their rare beauties by the axe of the feller, beneath whose strokes have disappeared some of the wildest and grandest oaks and other forest trees which could charm the eye of the poet or the painter. We looked, also, with much interest into a wild adjoining valley, called *Gobble's* Combe, which Hannah More, on our return, insisted was a corruption of *Goblin* Combe, adding, in support of her conjecture, that it was not far distant from *Failand*, or *Fairy-land*, another picturesque locality commanding fine views of the Severn. We found, also, the views around Barley Wood itself to be of the utmost beauty. It stands on a gentle acclivity, fringed on one side by natural woods, combining with flourishing plantations, which Hannah More herself had planned; and on the other side commanding exquisite views of an almost Arcadian valley, surmounted by a range of the Mendip Hills, and enlivened in the distance by a reach of the Bristol Channel, whilst the truly elegant tower of Wrington Church, rising above the village of that name, the birthplace of John Locke, imparted animation to the whole scene. Some tasteful root-houses, with appropriate inscriptions, graced the Barley Wood plantations, together with two monumental urns—one recording its having been erected at the expense of Mrs. Montagu to John Locke; the other, in affectionate memory of Hannah More's friend, Beilby Porteus, Bishop of London. Hannah More's conversation was worthy, in its point and acumen, of the friend of Dr. Johnson and of Edmund Burke. Speaking of a coarse, but clever

writer, she said: 'His sarcasms are inflicted, not with
a lancet, but with an oyster-knife.' In allusion to the
freedom of England from the sufferings inflicted
upon Germany and other parts of Europe, through
the ambition of the first Napoleon, she said: 'We
have only had a slight brush from the tail of that
terrible comet which has swept with such awful devastation over the Continent.' I asked her what was
the number of scholars in her various Somersetshire
schools? 'Had you asked me that question,' she
replied, 'some time back, I should have told you
they amounted to 1,300; but the infirmities of advancing age have obliged me and my sisters to put
them on the peace establishment, and they are now
reduced to 700.' She much gratified me before we
parted by expressing an earnest hope that my visit
to Barley Wood would often be renewed; adding, as
she pressed my hand, 'I trust that this day is the
commencement of a friendship that will last throughout eternity.'

For upwards of twenty years a most pleasant intercourse was kept up between us, and days spent at
Barley Wood were always days of intellectual delight
and religious interest. I had the pleasure of welcoming her and her sister Patty under my father's
roof at Blaise Castle in the autumn of 1810, when
she interested a circle of friends, assembled to meet
her there, by the charm of her wit and conversational

T

powers, no less than by her Christian philanthropy. As an instance of her sprightly good humour, I remember her whispering to me, as she looked at a youth of fat and florid, but vacant countenance, 'Can't you contrive to do some good to that young man who has so much of the corn, wine, and oil in his looks?'—adding, when I next saw her, 'How is our friend—Corn, Wine, and Oil?' After my marriage her affectionate kindness was shared by my wife no less than myself on all occasions, and we often dined and slept at Barley Wood, and also had the gratification of receiving her and Patty at our own residence. It was on one of these occasions that she described to me her first interview with Dr. Johnson. It took place at the house of Sir Joshua Reynolds, in Leicester Square. 'I felt,' she said, 'a little trepidation at entering the apartment where he was, in consequence of something Sir Joshua said to me just before we did so about his occasional roughness of manner; but he received me with a kindness which quite removed any such feeling. He was caressing a parrot, and on hearing my name he took my hand with the utmost cordiality, and addressed me in a few lines of one of my own poems.' Hannah More was interrupted in this narration by one of her zealous sisters, who exclaimed, 'Dr. Johnson has said that Hannah was the most powerful versificatrix in the English language.'

Such interruptions were not infrequent on the part of one or other of these devoted sisters. Hannah went on to tell me that on complaining to Dr. Johnson, how malevolently some of her writings had been attacked by the critics of the day, Johnson's sage reply was: 'Child, never mind them; don't you know that abuse is the second best thing? Praise may be the best; but oblivion is the evil.' Johnson delighted in her company, and when he was not in the best humour she had a way of parrying his thrusts which turned aside their edge, and provoked some pleasant sally which put all to rights again: she told me that she was once staying in Oxford when Johnson was there, and that she was invited to be present at a *fête* given in his honour by some of the members of his college, when the sage was in the highest good humour and delighted everyone. He was not a little attached to some of his Oxford friends, speaking of whom he said to her: 'Child, we were a nest of nightingales.' It was melancholy, she said, to see him when he was in one of his hypochondriac fits. She well remembered, on an occasion of this kind, sitting by him and hearing him, during much of an evening, repeatedly muttering the words in a low voice, '*God preserve me from the fear of death.*' She delighted, however, to dwell on the assurance that in the closing scenes of his life his feelings of dejection had passed away, under the firm but humble persuasion of his acceptance with God through the merits and mediation of his Divine Redeemer.* Mr. and Mrs. Garrick were among the most intimate friends of Hannah More when she lived in the world. Nothing could exceed their kindness to her; and her friendship with Mrs. Garrick, in particular, was warmly kept up till that lady's death in extreme old age. ' *All the Nine* ' was the familiar term applied to Hannah by Mr. Garrick; and she spoke with highest admiration of the wonderful powers of that great actor. An ardent admirer of those powers one day said to him, 'I wonder your feelings don't wear you out, for surely you cannot possibly give the expression that you do to tragic emotions without being deeply affected yourself.' Garrick replied, 'I am to play such a part (mentioning one of his celebrated characters) this evening, and if you will place yourself in such a box (mentioning which) you shall judge for yourself.' She did so, and she found herself opposite to Garrick in the midst of one of his most impassioned scenes:

---

* It will be found that this 'giant of modern literature,' in his last illness, humbled himself as a little child at the feet of Christ. Boswell says that Dr. Brocklesby, Johnson's physician, furnished him with the following statement:—'For some before his death, all his fears were calmed and absorbed by the prevalence of his faith, and his trust in the merits and propitiation of Jesus Christ.'—Boswell's Life of Johnson, vol. iv. p. 453. See also Hawkins's Life of Johnson, and Roberts' Life of Hannah More, vol. i. p. 377.

in the very acme of seeming emotion he looked at her for a moment, lolled out his tongue, and then went on as usual. It was in the garden of his villa at Hampton Court that a curious scene occurred between Hannah More and Lord Monboddo. That whimsical but learned man was a guest with Mr. and Mrs. Garrick at the same time as herself. They were walking together in the garden, when his lordship astonished his fair companion by a declaration of love and an offer of his hand and heart. His advances met with a positive refusal, and Lord Monboddo on this returned to the drawing-room, when he amused Mrs. Garrick not a little by telling her what had just occurred, adding, 'I am very sorry for this refusal: I should have so much liked to teach that nice girl Greek.'

Talking of Greek puts me in mind that Hannah More told me that Mrs. Carter, who gave such proof of her knowledge of that language by her translation of Epictetus, was so unpretending that you might have lived a twelvemonth in her society without suspecting her of such erudition. A Frenchman of literary celebrity, the Abbé Raynal, was at Bath, where he attracted much notice. It was well known that he avowed atheistic opinions. At a party in that city he was introduced to Dr. Johnson, and attempted to shake hands with him; but Johnson, in rejection of his advances, put his hands behind his back. Hannah More, on seeing him soon afterwards, said to him : 'Dr. Johnson, is it true that you were so rude as to refuse your hand to the Abbé Raynal?' 'Yes, child, it is true ; I won't shake hands with an atheist to please you or anybody else.' One of his inimitable sayings, as repeated to me by herself, is also mentioned by Boswell : 'How is it that Milton, who was so great a poet, did not write better sonnets?' 'Milton,' he replied, 'was a genius who could hew a Colossus out of a rock, but could not carve faces upon cherry-stones.'

She often dwelt with admiration on the lofty intellect and brilliant conversational powers of Edmund Burke, and used to tell of the delight which he imparted to her own evening coteries in Park Street, Bristol, in the intervals of his canvass for that city, of which, as it is well known, he was for some time the distinguished representative.

It was whilst she was in the midst of her literary fame and personal popularity that she was gradually brought under a deep conviction that to live to the glory of God and to the good of our fellow-creatures should be the great object of human existence, and is the only one which can bring peace at the last. Under this impression she quitted, in the prime of her days, the bright circles of the metropolis, and, retiring with her sisters into the neighbourhood of Bristol, she devoted herself to a life of active and

Christian benevolence, and to the composition of various works, having for their object the religious improvement both of the highest and humblest of mankind.

Having dwelt for some length on the literary charms of Hannah More's society, I must now add that her house and home formed a centre, where good men, and faithful zealous clergymen in particular, were sure to meet with a hearty welcome, and with true sympathy and support in their various efforts of a religious and benevolent nature. They were welcomed with the kindest hospitality, and left the house charmed and edified by the converse of its hostess, and grateful for the cordiality of the whole sisterhood. Her visitors were most numerous, many of them coming with letters of introduction from various friends. Lady visitors, in particular, were anxious to see and converse with so distinguished an ornament of their sex, whose writings, example, and influence had been promotive of the highest good throughout our country. The writings of Hannah More had been published and circulated extensively in America, and many were the instances in which travellers from that country visited Barley Wood to offer their grateful homage to its benevolent and distinguished owner. A constant stream of charity flowed forth from Hannah More for the relief of distress of all kinds. The tenderness of her feelings when acted upon by distress was most striking. The neighbouring district of Rodborough, inhabited chiefly by miners, who raised calamin, was occasionally visited with great want from the absence of demand for that article, when she not only relieved the suffering population, but bought up the calamin, selling it afterwards for the best price she could obtain. I well remember, on a pressing occasion of this description, when she was nearly eighty years of age, that after doing all that she could in this way, finding the suffering to be still severe, she could not speak of it without shedding tears.

Amongst the many public charities in which she took a deep interest, and to which she liberally contributed, none were higher in her esteem than the British and Foreign Bible Society and the Church Missionary Society, to each of which she left large bequests. In fact, the principal part of her fortune was appropriated to charitable legacies.

A not unfrequent visitor at Barley Wood was Tom, as he was familiarly called, afterwards Lord, Macaulay, who, in his boyish days, was greatly cherished by Hannah More. She was very intimate with his parents, and often have I heard her predict his future eminence, and dwell with lively interest upon his extraordinary acquirements and precocious talent. He, on his part, was much attached to his kind hostess; and I well remember with what delight he revisited Barley Wood in 1852, when Mrs. Harford and I conducted him thither, where he was most warmly welcomed by my brother, its present owner. The bright sisterhood had long since passed away. He talked much of them, and explored every part of the grounds with a pleasure and feeling strongly expressive of his attachment to their memory.

No visitor of learning and education could be long in Hannah More's society without feeling that she deserved to be regarded as one of the stars of that galaxy of talent which in the days of Johnson imparted so much celebrity to the literary society of London; though at the same time it was quite clear by her writings that her mind soared far above the desire of mere earthly fame. To do good to others had long become the leading principle of her life, and to attract the souls of those over whom she had any influence to the feet, and into the school of Christ, was, therefore, her daily aim. Everything of this kind was, however, done with simplicity of purpose and with no attempt at dictation; but she had a wonderful power of engrafting useful and excellent hints and remarks upon common topics.

Among various bishops of the day, Dr. Porteus, Bishop of London, and Dr. Burgess, Bishop of St. Davids, and afterwards Bishop of Salisbury, were in the number of her intimate friends and frequent correspondents. They were held by her in the highest esteem; so also was Dr. Shute Barrington, Bishop of Durham, who united to aristocratic dignity a most generous temper, and a heart deeply imbued with true Christian charity.

She once made us smile when speaking of some of her bishop-friends, by adding, 'I half incline sometimes to advocate, with Queen Elizabeth, the celibacy of the higher orders of the clergy. I have known so many good bishops spoilt by worldly wives.'

It was quite a treat to hear her read aloud either poetry or prose, and her expressive countenance adapted itself to the subject whatever it might be. Nothing could exceed her friendly good nature; and I remember, as an instance of it, that one day being about to take my leave of her while haymaking was going on at Barley Wood, and the servants all in the field, and the question arising who was to serve me with luncheon, she herself left the room, and in spite of my remonstrances brought in all that was requisite.

All interested in Hannah More have reason to feel grateful for the affectionate zeal with which Sir Thomas Acland, Bart., whom she most highly loved and esteemed, pressed her to give her consent to sit for her portrait for him; and having obtained it, he sent down Mr. Pickersgill, R.A., to Barley Wood for the purpose. It proved an excellent likeness of her in her old age; and an able engraving from it by Worthington has been extensively circulated throughout the country. The picture now hangs in one of the principal apartments at Killerton Park.

It was touching to visit Barley Wood after her four sisters had been consigned to their last home in Wrington churchyard. The kind and friendly voices, to the sound of which we had long been so familiar, had successively passed away; but Hannah More's Christian principle and strength of mind sustained and comforted her under these bereavements, though she could never allude to the death of Patty in particular without tender emotion. It so happened that Mr. Wilberforce was a guest at Barley Wood in 1819, when Patty was seized with an illness that speedily terminated in her death. After she became fully sensible of its approach, she opened her heart to him in a deeply interesting interview, the particulars of which he communicated to me on reaching Blaise Castle the following day. He described in forcible terms the patience with which he found her enduring great bodily suffering, and dwelt on her faith and hope in the prospect of death, through faith in her blessed Redeemer.

The following is an extract from a letter of my own, Oct. 3, 1819: —

' I went yesterday to Barley Wood. Mrs. Hannah More was at first much affected, but soon became composed. Nothing can be more resigned or more affectionate than the whole frame of her mind. She received me with unusual warmth of friendship, and soon broke out into many pious expressions of resignation as regarded her own loss, though daily felt to be most severe, and of full assurance of the blessedness of the change for Patty. She looks shattered and worn by the affliction; but, from all I saw, I think she will gradually rally again. You will recollect how devoted to her, in every sense of the word, Patty was, and how much she had made herself necessary to her by every species of affectionate and endearing attention: so that you can enter into her feelings when she says she has lost in Patty—hands, and feet, and eyes. Patty's last words were—I take them from Bishop Ryder's lips—" I love all the ways of the Lord, even my sufferings. I have done but little for God, and what I have, has been stained with imperfection and sin : but I think only of Jesus Christ and Him crucified." She was frequently repeating the first verse of the 27th Psalm.

'To promote the glory of God, to aid in diffusing both at home and abroad the influence of the Gospel, to improve and benefit her poor neighbours, to educate the young in religious principles, to comfort and relieve the destitute and afflicted, and to exercise habitual self-denial in order that she might more abound in works of charity—these had been the objects for which she lived and supremely cared.

'Her affections were peculiarly warm and tender, and won for her the fond attachment of her numerous friends. Her own profound humility contrasted strongly with their reverence and esteem. She clung by faith to the cross of Christ, and died as she had lived, a genuine Christian.'

To return to Hannah More. When she became thus bereft, her desire to do good and her zeal for the progress and promotion of Christianity furnished her mind with subjects of high and undying interest, and its elasticity never forsook her. A pious and estimable young friend became her companion, and assiduously ministered to her comfort, and various ladies with whom she was intimate were her frequent visitors. She finally moved to 4, Windsor Terrace, Clifton, in 1828, and there passed the remaining five years of her life. It may well be imagined that she could not quit Barley Wood without much regret, but various causes combined to render this step expedient. By moving to Clifton she was brought within the reach of many attached friends and excellent clergymen, amongst whom I may particularly mention the Rev. John Hensman, minister of Clifton, to whom she was greatly attached, and who discharged the office of Christian pastor to her with affectionate veneration. She was thus also brought near to Dr. Carrick, her able and kind physician, who had throughout more than one illness assiduously attended upon her at Barley Wood as a friend. It was also reported that domestic trials, proceeding from her servants, made a change of residence desirable. It was in allusion to this fact that she said to me, as I led her down stairs to the carriage which was to convey her to Clifton: 'Like Eve, I quit paradise; but not, like her, driven out by angels.'

During her residence at Clifton we were often with her; and in the course of two visits from Mr. and Mrs. Wilberforce, in 1831 and 1832, we had the pleasure of taking them to call upon her. On the first of these occasions we had a most animated interview with her. She traversed her apartment backward and forward, leaning on his and my arm, and talked with us in a strain of Christian piety, mingled with playful and bright ideas. On the second visit referred to, it was unfortunately a day of extreme debility, and to see Hannah More otherwise than lively and animated was equally novel and painful to her friends; but she gently rallied at the sight of Mr. Wilberforce, and he laid himself out to amuse and interest her. We carried away with us a strong impression that her mortal pilgrimage was fast approaching to its close, and so it proved; yet, though in her 88th year, she outlived by a few weeks the kind and beloved friend who now, to all human appearance, promised to be her survivor.

The impression we had thus received induced us to visit her more frequently, and the following are memoranda of a visit which Mrs. Harford and her sister paid her in August 1832. They found her looking well, and with undimmed brilliancy of eye. Her memory was occasionally clouded; but the affections were in full play, and love beamed in her countenance. She told them a favourite motto of hers was *faire des heureux*; and it may be truly said that she was actively at work in this way throughout life. As she paced her rooms, leaning on their arms, she called on them to admire the beauty of the views which the windows commanded. Mrs. H. said *it was a pleasant cage*; and her sister added, '*It contains a Bird of Paradise.*' She would scarcely let them go, saying several times over—
' Will you come again? Will you come soon?'

In November of the same year (1832) I was with her, and I will transcribe from my notes a few of the expressions which fell from her lips as we conversed together:—'You behold a dying creature: pray for me. I cannot describe the comfort and support I find from these words, "I *know* that my Redeemer liveth;" not I believe, or I hope merely, but *I know*. What confidence! I live on the Psalms. I have lived a long life. I might have lived a more useful one; but I have a gracious Saviour, and His peace is in my heart. I feel His presence with me.' I repeated to her various consoling passages of Scripture, such as the 23rd Psalm, the closing verses of Romans viii. &c. &c. She responded with deep feeling to each, adding, how invaluable is a Christian friend.

In one of our visits to her in 1833 we found her seated in an arm-chair, with a Bible open before her, waiting for the arrival of the Rev. Mr. Hensman, who was coming to administer to her the Holy Communion. She was in a sweet and a most Christian state of mind, and received us with warm affection.

The last nine months of her life were marked by a great decay of her powers, both mental and physical; and those who were about her beheld only the wreck of her former self: yet the heart never ceased to glow with affection.

Only a short time before her death, as I stood by her bed-side, I took her hand and kissed it. She opened her eyes, and drawing my hand to her lips kissed it more than once, and then stretched out both her arms towards me. How far she really recognised me could not be ascertained; but it seemed evident that she was conscious of the presence of a dear friend, and was herself under the influence of that *spirit of love* which so strongly marked her intercourse with all who were dear to her.

The day before her departure I paid her a brief visit with one whom she had most highly loved and esteemed through a long course of years—Sir Robert Inglis. We stood by her bed, but she was too far gone to recognise us. We watched her for some time with deep and affectionate interest, and reflected on the bright exchange she was about to make from a bed of suffering to the glorious presence of her Divine Redeemer.

She entered into her rest on Saturday, September 7, 1833, in her 89th year. I called at the house that day, but the happy spirit had been just set free. We deeply felt the severing of this tie. How much do we owe, and have we owed, to her invaluable friendship and affection—as well as to that of the beloved friend, Mr. Wilberforce, who so shortly preceded her to the eternal world.

She was very fond of flowers, and we had often sent her nosegays from Blaise Castle. She had once said she should like some to be strewn over her when she was gone. It was Mrs. H.'s privilege, and that of her sister, when, on September 9, they visited her remains in their last abode, to realise this wish. It may be imagined what tender and impressive recollections arose in their minds while in this solemn chamber.

The two following letters will appropriately close the preceding brief tribute to the memory of Hannah More; the last of which, addressed to a family in deep affliction, is now for the first time printed.

'Shrewsbury: September 9, 1811.

'My dear Friend,—Accept a hasty line for your entertaining letter. I have been so constantly in motion, or in company, or indisposed, that I have not written one letter but of absolute necessity, or business, since I met you that last morning. You have not the less lived in my affectionate remembrance. Instead of the stipulated fortnight, Mr. Gisborne detained us a month in his charming forest, accompanying us, however, on our excursions. We obeyed your commands in making the Derbyshire tour. Matlock is enchanting, of a different character but not more interesting than Malvern, where we stayed a couple of days in our way to Staffordshire. Everything concurred to make our visit at Yoxall interesting; scenery of a peculiar character, and pleasant society in the house and neighbourhood, in which are several truly excellent clergymen, among whom is Mr. Cooper, author of the valuable sermons. Among our inmates was Mr. Ryder, brother to Lord Harrowby, the bent of whose mind, and the turn of whose conversation, incline me to believe that he is not unworthy to fill the pulpit at Lutterworth, once so worthily filled by

Wickliffe. It is delightful to witness the many accessions to the cause of Christian piety in the higher ranks of life. We are come to this fine old town to visit some friends. Both the near and distant views are intimately connected with our history. Here is Battlefield, where Harold once fought; and since still more distinguished by the fall of Hotspur, Harry Percy: they do not exactly show the spot where *Falstaff ran away.* Another hill presents the scene of the valour of Caractacus. Another of an ancient oak, said to have been planted by Owen Glendower. Still more substantially valuable are the numerous edifices consecrated to public charity; all appear to be remarkably well conducted. With public charity the name of Richard Reynolds naturally connects itself, as it did in Coalbrookdale, the most wonderful mixture of Elysium and Tartarus my eyes ever beheld. Steam engines, mills, wheels, forges, fires, the dunnest and the densest smoke, stupendous iron bridge, all rising amidst hills that in natural beauty rival Dovedale and Matlock. We grieved that excessive fatigue and heat, rendered more intolerable by a withering east wind, prevented us from roving through Reynolds's fine walk, which he kept up for the benevolent accommodation of others. To-morrow (alas! it is still a parching east wind) we purpose, if it please God, to set out on a little Welsh tour with our hosts, to peep at the Vale of Llangollen, Valle Crucis, Chirks Castle, &c. &c. We hope to return over the classic ground of Ludlow, a town I much wish to see. May God bless and direct you, my dear friend.

'Yours affectionately,
'H. MORE.'

'My dear afflicted Friends,—What can I say, but that I mourn with you from the bottom of my heart? Yet, in the midst of sorrow, I must and do rejoice that you have the best of all consolations, that of knowing that the dear departed is a saint in glory. I was so much affected with your letter that I was not able to write to you last night. After having read it three times, I despatched it to ——. We prayed for you all in our family devotions, and read Venn's beautiful sermon on the happiness of heaven, and closed with the fine funeral hymn. It was a solemn evening. I have comfort in thinking that this heavy blow has fallen on a Christian family. If calamities are so deeply (perhaps the most deeply) felt by the truly pious, I have often wondered how those who have no support from Divine grace bear them at all.

'How I envy the sublime dying scene of this young Christian. She had everything to quit which renders life dear and pleasant—youth, beauty, rank, fortune, an affectionate husband, lovely children, kind relatives! The way before her seemed peculiarly smooth, life was all promise, and the prospect presented nothing but hope. She is called on to abandon all this, called unexpectedly indeed, but, as she has evidently shown, not unprepared.

'Oh! the mysteries of the Divine dispensations. God's ways are not as our ways. *She* taken! *I* left! She cut off in the midst of her usefulness; I spared, after the little I could ever pretend to is past. God saw that in a short time she had fulfilled a long time, and graciously gave the crown almost before she had run the race.

'I had just been writing, on an occasion very similar, to the ———. Their amiable daughter, Mrs. ———, is left at the age of thirty with seven little orphans. She is a most exemplary young woman, and here again is seen the goodness of God in causing this heavy trial to fall on a family of great piety, who are supported by those consolations which are neither few nor small. It was a comfort to learn that ——— was on the spot. She has taken many hard lessons in the school of affliction, but she will bear in mind *who* it is that the Lord chasteneth. I hope that dear ———'s feeling heart and delicate frame will not cause her to suffer in her health. Pray assure the whole mourning party of my cordial sympathy. It is all that impotent friendship can offer; it can feel, but it cannot serve. With many thanks for your so kindly writing to me, I remain, my dear friend, yours sincerely,

'H. MORE.'

CHAPTER XIII.

Sketch of the Life and Character of the Rev. R. C. Whalley, B.D., Rector of Chelwood.

WE here introduce the promised particulars of the character of the Rev. Richard Chapple Whalley, referred to in Chapter II. His father was the Rev. John Whalley, D.D., Master of Peter House, Cambridge, and Regius Professor of Divinity in that University. His mother was a daughter of Francis Squire, Chancellor of Wells. He was born in the year 1748. His father died while he was an infant. Where he was educated I have not been able to ascertain; but his attainments as a scholar were such as to prove that it must have been at some superior seminary. A Latin inscription from his own pen, on a tomb in Wells Cathedral, in memory of his wife and of his son Francis, a boy of great promise, who prematurely died at Eton, proves that he composed in that language with point and elegance.

At the time that he naturally would have gone to a university, such was his wish to devote himself to the study of the Fine Arts, that he went for this purpose to Italy, where he resided for several years, with the view of making Art his profession; but his health failed under the needful application, and he returned home without having any settled plan of life. Unhappily, during his residence in Italy he became familiar with the writings of Rousseau and Voltaire, and also with the French philosophy which at that time greatly pervaded the Continent. His religious principles for the moment were much shaken; and he returned home with opinions, tastes, and habits uncongenial with those of the friends and associates among whom he was to live. The elegance of his manners, his refined taste, and his familiarity with foreign scenes and objects, naturally made his company sought after; but there was a hauteur and reserve about him which was justly deemed repulsive. His elder brother, the late Dr. Whalley, once said to me: 'You are well acquainted, I believe, with my brother Richard; but, unless you could compare him, as I can, with his former self, you can scarcely imagine what religion has done for him. I well remember him as one of the proudest and most fastidious of human beings. He was, in fact, the proudest man I have ever known.'

The late Mrs. Hannah More and her sisters, who were well acquainted with him at the time now referred to, have occasionally described him to me as a man of refined elegance but chilling fastidiousness. Scepticism, when it is based on presumptuous assumption, is in the same proportion arrogant and uncompromising. Voltaire, with all his wit and powers of ridicule, was, in this sense, grossly ignorant, as has often been proved; yet he touches on theological topics with the confidence of an oracle: and from scepticism, in all probability, originated much of that peculiar pride which marked Mr. Whalley, and which cost him in the retrospect the sincerest grief and self-abasement. To the absence of religious control is also to be ascribed the too great dominion that a naturally hot temper had obtained over him, and which singularly contrasted with the benignity and meekness of his subsequent demeanour.

Amongst the friends who lamented his sceptical opinions there was one who not only urged upon him the duty of farther enquiry, but entreated him to bring those opinions to the test of such a book as 'Butler's Analogy.' Happily for his own future peace, he followed this friend's advice, and devoted himself to its perusal. It met and refuted the arguments upon which he had most confidently relied as subversive of Revelation, and on laying down this book he cast away his sceptical opinions—he did homage to the truth of the Bible—he saw through and rejected the sophistries to which he had yielded credence—and avowed himself a firm believer in the

Christian Revelation. These were important points gained, and the change in his opinions wrought a beneficial change also, in many respects, in his tastes, pursuits, and habits. But though his judgment and reason were convinced, his heart was not yet duly affected. He was too much inclined to view the subject of his late enquiry as a dry question of truth and error, without a proper regard to the weighty consequences and obligations consequent on his faith as a Christian. He, therefore, felt no depth of remorse for years unprofitably spent, for talents misapplied, and for mercies abused.

In 1775 he married Elizabeth Frances, youngest daughter of the Rev. John Paine, Canon of Wells. It was a match of mutual attachment, and imparted to him, during twenty years, much domestic felicity. Two children were the fruit of this union.

In 1786 he revisited the Continent with his wife and children, and during his stay there he was induced to take orders, that he might hold a living for his nephew, a minor; and leaving his family at Tours he returned home, and was ordained in June 1787.

He entered on his important office with the intention of discharging its functions in a respectable and conscientious manner; but when, at a subsequent period, he became deeply impressed with the end and the object of the Gospel, he felt that he had fallen far short of the high and holy principles essential for the faithful discharge of the duties of the Christian ministry.

In 1795 he lost his beloved wife, the idol of his affections; and such was the affliction into which he was thrown by this calamity that life seemed to have no charm left for him, and he fell into a state of confirmed dejection, from which nothing seemed able to arouse him. In this state of mind he quitted home for a twelvemonth, with no companion but the Bible, and sought for consolation in its daily perusal; but at the end of that time he returned home much in the same state of mind in which he left it. In what way he at length was relieved from this heavy burden and found peace with God will be best explained in his own words in the following passage from one of his letters. The commencement of this letter has been lost; but there is no doubt, from the part which I am about to quote, that it described to his friend his former self as having been vain, worldly, and irreligious. He then proceeds as follows:—' Of these things, and what I might have done to benefit the soul of one whom I thought I loved as my own soul, I have the bitterest feelings of remorse, and can hardly restrain my agonies; but God is all-wise and all-sufficient to all, and orders all things wisely, for His will is wisdom. I rest in that persuasion, and well know that Christ was able to make Himself known without my concurrence. Such, dear madam, is the picture of one who passed with the world for a respectable—nay, and with many, I believe—for an exemplary character. In this way are the eyes of men deceived and cheated. But God, who had determined to show mercy to one who was so peculiarly unworthy of mercy as I was and am, was preparing the instruments of an entire conversion and change of my heart and soul. Severe and sharp were the means, it is true, and such as then appeared in the shape of the deepest afflictions; but nothing less than caustics will avail when the gangrene is far gone. He who could not be prevailed upon by mercies and blessings was touched and awakened by sufferings and punishments; I lost my child—I lost my wife—I had many pecuniary losses and crosses. The consequences of extreme vanity and mismanagement in my temporal affairs began to press upon me; and when, after twelve months' absence from the scene of my affliction, I returned to my parish, of which I thought I had determined to take leave for ever, I found all these recoil with inexpressible weight upon me. Left entirely to reflect on myself, and at full leisure to estimate the extent of the troubles God had brought upon me, and those He had suffered me to bring upon myself—finding myself despoiled of all I thought was essential to my comfort and happiness, and harassed and embarrassed with a thousand domestic inconveniences and difficulties—I began to look for comfort to my religious principles; but now, alas! to my unspeakable confusion, found that they were fancied principles, and that I really had cultivated none that could effectually help me. I had given myself, in the hour of careless security, credit for a faith that was too light and unsettled to bear me up in the hour of difficulty and distress, and therefore I had nothing to comfort me—I had no place to flee unto—and no man cared for my soul. In this state and condition of misery, with the Bible one day in my hands (for I had daily resorted to it for consolation since my calamity, though I now seemed not to have found it), I uttered, in the bitterness of my heart, a prayer to God, that if that book were indeed the word of truth and message of salvation, He would enable me to apply to it with a simple, teachable, unbiassed mind. I offered a supplication to my Saviour Christ, that if He were indeed a Saviour, He would be at hand to save me; that He would help my prejudices, and heal my corruptions, and remove my blindness, and make me see the wondrous things of His law; and enable me to understand what was there revealed, and give me strength to believe in it and apply it to my own emergencies: for well I knew that faith in the Scriptures would be a cure for all the evils I laboured under; and that other cure there was and could be none. My prayer was offered under a complete sensation of my inability to help myself in the removal of them—a perfect persuasion that if God gave no comfort, I could never more have comfort, either in time or eternity. Blessed be God that He heard my prayer and turned not aside His mercy from me. I soon—nay, almost instantly— began to read the Scriptures with new eyes. I almost instantly perceived that they were the true Word of God. But, gracious God! what did that perception bring with it? A perception derived from that Word which told me I was born in sin—a perception of a life spent in actual sin. I now more than suspected that I had been hitherto blind, and that all I had hitherto done was madness and folly. I now more than suspected I had misused every talent, and abused the long-suffering and goodness of my God, and had been adverse to Him in every particular and passage of my life. The things I had hitherto looked upon as mere follies were now felt to be the rankest sins; and I saw at once, and with the utmost clearness and precision, the sin of self-will, and self-love, and love of the world—that my whole will and mind were sinful—that from a bitter fountain sweet waters could never have flowed for one moment—and that, therefore, I had naturally never done one thing that could be acceptable to God. The dreadful consequences of remaining under the wrath of God occurred to me with a tremendous force. I will not lengthen this account by particulars of my sufferings; but I dare say you need not now be told that I began to understand the scheme of Justification by Faith alone, and that what I had never comprehended or cared about before appeared the plainest and simplest thing in the world to me.'

Brought to this temper of mind, Mr. Whalley was now well prepared to receive the glad tidings of the Gospel. He deeply felt his own sinfulness and the need and value of a Saviour; and, therefore, the great mystery of Redemption became the subject of his joyful, thankful, and adoring contemplation. He was thus led to enlightened views of the Christian dispensation. An inward calm—a heavenly repose of soul—a filial confidence—a settled hope in the mercy of God through our Lord Jesus Christ—a sense of reconciliation now quieted the tempest of his feelings. His soul became full of serenity. It rejoiced in the clear sunshine of Divine love. By day, by night, this glorious subject absorbed his thoughts and fixed his contemplation.

Similar instances of the power of Christianity in thus renovating the human heart are of perpetual occurrence, and form important internal evidence of its Divine origin. The change of character thus produced in individuals is often so great as to strike those with wonder who, from want of due acquaintance with the Bible, are not aware that effects such as these are in perfect accordance with the promises of God, addressed in its sacred pages to those who repent and believe. To know Jesus Christ as the Saviour of our souls we must have grace to believe in Him; to advance and persevere in holy obedience to His will, we must have vital union with Him. The doctrines of Grace are, it is true, capable of dangerous perversion; but our Saviour has furnished an infallible test, applicable to all cases of alleged conversion, 'By their fruits ye shall know them : men do not gather grapes of thorns, nor figs of thistles.' Tried by this test, the great change of sentiment and feeling experienced by Mr. Whalley, may be safely designated as a remarkable interposition of Divine mercy, for, while it opened his eyes to the one only basis of a sinner's justification before God, it rendered him victorious over his habitual sins and corruptions—it imparted to him elevation of purpose and action—it filled him with supreme love to God and brotherly kindness to man. Henceforth a great and marked alteration was visible, not only in the tone and temper of his mind, but also in his doctrinal statements, and in his assiduous attention to his pastoral duties. The heat of his natural temper was controlled and subdued—the pride and fastidiousness of mind and manner, of which his associates had so often complained, was succeeded by humility, forbearance, and love. The contrast presented by his present and past character was so striking, that it naturally became a theme of conversation amongst his former acquaintances and friends; and many of them could scarcely believe that the amiable and saint-like man, whom they now viewed with reverence, was the elegant and fastidious individual who had formerly seemed scarcely accessible. The topics and the tendency of his preaching assumed a totally new character. As was said of Richard Baxter's addresses from the pulpit, 'he spoke as a dying man to dying men.' Perhaps its leading characteristics could not be better explained than by stating that it was in the spirit of that striking passage in one of Bishop Horsley's charges, in which he exhorts his clergy as follows:—

'Apply yourselves with the whole strength and power of your minds to do the work of evangelists. Proclaim to those who are at enmity with God, and children of wrath, the glad tidings of Christ's pacification; sound the alarm to awaken to a life of righteousness a world lost and dead in trespasses and sins; lift aloft the blazing torch of Revelation, to scatter its rays over them that sit in darkness and the shadow of death, and guide the footsteps of the benighted wanderer into the paths of life and peace.'

Mr. Whalley thus became the affectionate pastor, the enlightened teacher of his flock. He was not only prompt to render them effectual help, to the best of his ability, at the call of sickness, or poverty, or sorrow; but he made himself personally acquainted with their various characters, and, as far as possible, with their religious condition. From day to day he watched over them for good; and to his exhortations and active endeavours for their improvement and happiness he united earnest prayers to God for a special blessing on his labours.

A sacramental prayer is given as a specimen of his devotional feelings at this time :—

'Lord! give me grace, a miserable sinner, to lay hold on Thy mercies in Jesus Christ. Lord, I believe: help Thou mine unbelief. Strengthen my faith in the sacrifice I am about to commemorate, and teach me, while I fly to it from the terrors of a wounded conscience, to make it the assured pledge of my redemption, by a renewed and sanctified conversation for the time to come. Oh! let it ever be a means and a motive with me to an effectual change of heart. Let it gradually work in me some faint resemblance of Thine adorable perfections—a pious submission to Thy blessed will—a patient conformity to Thy most holy life—a spirit inclined and qualified to glorify God on high, and to follow peace on earth and good-will towards men. Grant, O merciful Father, that the world may be crucified to me, and I unto the world, through Him who loved me and gave Himself for me, Jesus Christ the righteous. Amen.'

In the year 1800 Mr. Whalley gave up the living of Horsington to his nephew, and soon after took possession of Chelwood, a small rectory near Pensford, in Somersetshire, given him by his old friend and connection, Dr. Beadon, Bishop of Bath and Wells. The population was chiefly rural, and the kindest feelings quickly sprang up between them and their pastor. The daily proofs which they received of his disinterestedness and heartfelt zeal for their present comfort and their eternal welfare secured to him their reverence and affection. Even those who refused to profit by his warnings could not but appreciate his character and example. His Christian tenderness towards the youthful members of his flock was truly impressive. At a later period than that now referred to an instance came under my own notice in which he watched, like a ministering angel, at the sick couch of a young person who was dying of consumption, and to whom his ministry had been eminently useful. In the last stage of her trying illness he even received her and one of her relatives into his own house for change of air and better attendance; and he was finally rewarded for his great kindness, not only by her heartfelt gratitude, but by witnessing the perfect peace and joy with which, through faith in the Redeemer, she breathed her last.

He was fond of resorting to Barley Wood, the residence of Mrs. Hannah More. No one more admired Mr. Whalley in his new character. She was used to say: 'A great many good men are *near* Heaven, but Richard Whalley is *in* Heaven.'

In the year 1810 a friend of Mr. Whalley's, who was an East India Director, lost his wife, to whom he was tenderly attached. He wrote to his old friend, claiming his sympathy, and anxiously enquiring if he thought that he should recognise in a future state the lost partner of his affections. The question touched Mr. Whalley's feelings on a tender point, but he did not allow himself to be drawn into the discussion of it, as he had reason to think he should be more truly acting the part of a faithful Christian by entreating his correspondent rather to test the validity of his own title to the heavenly inheritance than to speculate on such a topic. The sound judgment and affectionate sympathy with which he pressed this delicate point on his friend's attention were such that his disappointment at not being indulged with a letter in the strain that he had expected was forgotten in his admiration of the wisdom and sincerity with which Mr. Whalley had addressed him. Under this impression he transmitted the letter he had received to his friend Mr. Wilberforce, at that time M.P. for the county of York. The letter was returned with the following reply:—

'Herstmonceux, near Battle: July 26, 1811.

'Many thanks, my dear sir, for your friendly communication, and for the truly Christian epistle you have entrusted to me. It is a happiness—a happiness for which, I doubt not, you are duly thankful—that you can call such a man your friend. Mr. Whalley's former letter, which you were so obliging as to lend me last year, I returned, and a precious letter it was. In sending back that which I have now before me, I will add a few words, which the perusal of it suggests, because they are of a sort congenial with this day's purposes and objects:—

'Your good friend, my dear sir, truly says that the religion which I approve, and which I wish to possess, is such as he recommends to you—the religion of the Bible, of the holy apostles, of the pillars of the Church of England, and of your most excellent and venerated friend, Swartz. Holiness of heart and life, produced by the operation of the Spirit of God in those who have been accepted through Christ Jesus, the Redeemer; that is, having repented and believed on Him, they have become sons of God, according to the expression of St. John i. 12: "As many as received Him, to them gave He power to become the sons of God; who are born (as the next verse says) of God;" *i. e.* by His converting and sanctifying Spirit. This, as your correspondent says, is what the natural man receiveth not. So St. Paul says before him (1 Cor. ii. 14): "It is foolishness unto him, because it is spiritually discerned." And wherever it really exists, there certainly is a disposition not to be conformed to this world; not, however, from gloom or austerity, or from any idea that God is pleased by our denying ourselves what would give us pleasure, and subjecting ourselves to pain, but because the true Christian finds from experience that worldly pleasures have a powerful tendency to blunt the edge of his spiritual appetite, and to diffuse a fog—if I may so express myself—through his mind, so as to cause spiritual objects to be seen less distinctly, their beauties and excellences to be less clearly recognised, and consequently less keenly relished. The truth is he endeavours to retain a practical impression of that astonishing promise—" Come out from among them, and be ye separate, saith the Lord, and I will receive you, and will be a father unto you, and ye shall be my sons and daughters, saith the Lord Almighty" (2 Cor. vi. 17, 18). Remembering this gracious declaration, and remembering also the apostle's injunction, twice repeated, not to "grieve," not to "quench" the Spirit, he avoids not merely that which is absolutely unlawful and positively forbidden, but whatever he conceives would be unsuited to his blessed inmate —whatever may dull or deaden, not merely whatever would extinguish, the flame.

'O, my dear sir, I talk and write of these things

fluently; but how much harder is it to preserve a lively sense of them in their reality and power. And therefore, how do we need all that can stimulate and quicken, rather than subject ourselves to the action of causes which have a directly opposite effect! But surely it can be scarcely necessary for me to add that all this inward religion is not to terminate within; but having there its seat—*there* being the living fire—thence, as from the sun, the centre of the system of our universe, will be diffused those quickening beams which are known through the world as the producers of beauty and fertility. In like manner these peculiar people are zealous of good works, and are only grieved that they cannot still more glorify their gracious Father and Redeemer, and benefit their fellow-creatures. May you and I, my dear sir, and all whom we hold most dear, experience more and more of the Divine transformation of which I have been speaking. Shall I express myself in Scriptural language, and say: "May our God fulfill in us all the good pleasure of His goodness, and the work of faith with power?" (2 Thess. i. 11.) May we find ourselves growing in grace, that we may hereafter partake of a more abundant measure of glory. *We may.* For assuredly the nature of God and of Christ is unchangeable—the same yesterday, to-day, and for ever. And when we consider what a prize is before us, surely it might be expected that our most vigorous efforts would be called forth; but, alas! here we are apt to be lukewarm and remiss—here, where our everlasting interests are in question! Quicken us, O Lord, by Thy heavenly grace, and let us not sleep as do others, but resolutely set ourselves to run, to fight, to *agonise* (as the Greek expresses it), so as not to run uncertainly. Above all, let us not flatter ourselves that we have attained, until this is established by clear decisive evidence. My dear sir, I am persuaded I need make no apologies for the freedom with which I have written to you. Adieu! and as you Orientalists phrase it, what can I say more?

'Ever sincerely yours,

'W. WILBERFORCE.'

It was often my wish to have made Mr. Wilberforce acquainted with Mr. Whalley. I well knew, from intimate friendship with both, what pleasure they would have felt in each other's society; but the opportunity never presented itself.

When Mr. Whalley's health allowed him to receive visits from his intimate friends, nothing could exceed the cordiality and kindness of the welcome which awaited them. It always appeared to myself, when I enjoyed this intercourse, that I had been admitted to converse with a sort of human angel. His powers as a letter-writer were of no ordinary character. They chiefly turned on religious subjects, and were greatly prized by those who possessed the privilege of his correspondence. As a specimen, for instance, of the way in which he could unite comfort and edification in addressing those who were in affliction, we subjoin the following letter, written by him to a gentleman who had very recently lost his brother:—

'Chelwood: January 11, 1812.

'My dear Friend,—Being now got home, and in a situation to reflect without interruption or distraction upon the occurrences of the past week, I find I have great reason to be thankful for them, and am, indeed, full of comfort. Heavy as the dispensation which has taken away your dear brother from us is (I mean to our weak flesh and natural affections), how can we do otherwise than rejoice at the thought of the glory that is now revealed to him, and the heavenly rest he has arrived at? It has pleased the Father of mercies to shorten *his* pilgrimage, and rescue him early from a world "that lieth in wickedness," while *we* have still to struggle with it: *his* warfare is accomplished, *ours* continues. Who that from above looks down (if such a view be permitted) upon the *happiest* lot of the righteous here below but must pity them in comparison with him who is now altogether and *every way* righteous—who has *found* that state of sanctified perfection which he was breathing after here, and has *for ever* found it. Had he still lived among us, he would, doubtless, have been favoured, as we are, with seasons of sunshine and cordials of grace; but the everflowing *rivers* of pleasure and everlasting sunshine are not to be enjoyed but where he is. There no cloud shall ever again interpose between him and his Redeemer—his sun shall no more go down, and the days of his mourning are ended. Therefore, my dear friends (for I speak here to your whole family), let *us* not mourn, but bless and adore that God whose ways are all perfect, and cheerfully address ourselves to accomplish the things He would have done by us, till *our* time shall also come. Only let us all take care that the one great thing be accomplished—the one great work of God—*the work of believing* in His Son Jesus Christ—the work of coming to God in His own way, and then walking with Him in His own paths, and by His own rule; the work of casting ourselves upon the Saviour's merits, and washing ourselves in the Saviour's blood, receiving and trusting to the atonement and to nothing else, and yet living and acting as if our own penitential humiliations and renewed state of obedience were our only security. In this manner did *he* receive the atonement, and in this manner let us receive it, rejoicing in the hope of the glory of God, and having the love of God shed abroad in our hearts by the Holy Spirit, as a testimony that our hope is such as will never make us ashamed. "This," said he, as we were listening to his words of Divine peace and love, "this is the joy of the Spirit;" and can we doubt it was so? You and I, who were witnesses to it, and to the tide of hope and affection that rose in his heart at the time, and flowed out from his lips, can have no doubt about the source from whence it came. Nay, our own hearts accorded to it, and told us plainly that "there is one body, and one spirit, and one hope of our calling, one Lord, one faith, one baptism, one God and Father of all, who is above all, and through all, and in us all." Let us, then, be thankful that the seal was thus sensibly set to his soul that God is true and faithful to His promises. That this God is our God for ever and ever, and will be our guide unto death, and that this precious soul was so favourably and with so little pain disentangled from the body, is another cause of thankfulness, though now, indeed, of no consequence, except to put us upon reflecting how weak we are after all in laying any very great stress upon circumstances of this sort (which are but for a moment, whether easy or painful), when we know them to be the prelude to an "eternal weight of glory."

'I hope, my dear friend, I shall soon hear from you that you are all in a state of peaceful resignation under this trial. The book he designed for me I shall consider as a treasure, which indeed it is. As to anything else, under all the circumstances of the case, I would wish to decline it. "Is it a time to receive money, and garments, and olive-yards, and vineyards," &c. at the risk of any imputation of our simplicity and sincerity in doing God's *work*, when every eye is upon us, and every soul is as it were interested in our *disinterestedness*. I have a thousand times wished that I had a moderate subsistence of my *own*, that I might feed the flock of Christ without remuneration. Yet even *here*, perhaps, there is a lurking pride, so that we hardly know what to wish or pray for as we ought, unless in that broad comprehensive prayer, to be inclined to that which will most *honour God and edify one another*.

'Lastly, my dear friend, let me beg you to seize upon this occasion to renew your resolution to lean at *least* but lightly upon the things of time and sense, and not to expect more from them than they are capable of, for what can they do for us, independent of the enjoyment of God? The very best love of the creature is but the shadow of His love; and how great is the danger that while we are pursuing the shadow, we may lose the substance! Yet we must *act*—yet we must form relations, and both taste the pleasures and do the duties that arise from them; only let us remember the end, the short con-

tinuance of all these things; and let every instance of mortality that is brought home to us, like the present, abide in our minds, to moderate everything that is excessive there. May God keep you from all evil, and bless you with all good, is the prayer of
'Your faithful Friend.'

The following is an extract from the letter which, as already stated, was so highly valued by Mr. Wilberforce:—

'Towards our walking with God, then, I have submitted to you three things—An habitual recollection of our high calling as His very children—an habitual recollection of the astonishing way by which we have been thus adopted to Him—and an habitual use of the means of keeping such recollection alive in us. But I am aware you will say that these things are rather the very substance of what we are enquiring about than a mere method of getting at it. Still it remains a question how these habits are to be wrought in us. How are we thus to remember whose we are and whom we serve? How are we thus constantly to bear in mind that the Lord Jesus made Himself a servant for us? How are we to have the constant mind and determination to draw nigh to God in all the *means* of recollecting Him? The world, with all its tinsel, will be dazzling and debauching us. The devil, with all his *chicaneries*, will be practising against us; and our flesh and nature will be continually giving way to solicitations and proposals so congenial to a secret corruption there. What is there, it may be asked, to be further done, or rather first done, to lead to this? *Pray without ceasing*, to draw down the graces and gifts of the Spirit into the soul, *and then go forth and practise them*, that you may *know* you have God near you; *for God meeteth him that rejoiceth and worketh righteousness, those that remember Him in their ways.* If we consider that prayer is an actual approach to God in a way of holy familiarity, to consult with Him about our souls, and carry on a business of infinite importance with Him; a way in which we seek Him for that purpose with all the accelerated motions that His own grace can inspire us with; if we consider that we are drawn by His own blessed Spirit in the exercise of prayer, and furnished with arguments, and filled with fervent desires, "the Spirit itself making intercession for us with groanings which cannot be uttered;" if we consider by whom we are taken in hand and introduced into the holy presence in prayer, and whose blood we plead, as well as whose Spirit it is that enables us to plead it, we must perceive, I think, how much our conversation and habitual walk with God must depend upon prayer, and that they who are most frequent and most fervent in the exercise of it will be most nigh to God, and the greatest proficients in walking with Him. If the truth be, as it certainly is, that the highest and closest act of communion with God is by prayer, and that all the graces of the Spirit are strengthened and increased by it: if we look upon the Father as it were in this manner through Christ, the more we thus look upon Him the more we shall remember Him we so looked upon—the more we shall carry away with us into our commonest employments His blessed image—the more will everything we see and do bear some impression of it—the more will the taste and relish of His grace and goodness remain upon our palates, till we come to that feast of prayer again, which will make all those things insipid and uninteresting to us that would otherwise have drawn us aside from Him. But now, if we have this grace abundant in us through prayer, if the face of God shines upon us in it, and the taste of His love be prevalent, then let us go forth and make this derived and communicated light shine before men. Then let us make this love of His to us the model and the means of our love to others—of such a love as delights in doing good, and rejoices in the endeavour to diffuse joy and gladness around us, and especially the joy and peace that belong to right religion. But, indeed, every act of charity that is done upon the principle of God's love to us, must necessarily keep Him nigh us, and make

Him present with us. We must be up and doing for God—we must work and labour in this way—if we would have Him come to us and encourage us, and make us glad with His countenance, and give us cordials of refreshment. As I said before that the master's presence will animate the servants in their work, so the servants who work well and diligently may expect that their master will have a pleasure in being with them. When the pleasure of the Lord prospers in their hands, and He sees His designs accomplishing, it is not likely He should be long absent from them. At least, my experience warrants me in saying this. When I labour most (though little is my labour at the best), I feel God nigh unto me; when I remember Him in His ways, He seems to remember me; when I rejoice to do His work, He meets me in it; when I am sluggish, 'I find Him not on the right hand, and on the left He hideth Himself that I cannot see Him;' but in the humblest and commonest work I see Him most, when it is offered to Him. I therefore think (for I am sure it is time I should release you) that the sure way of *setting God before us and of seeing Him that would otherwise be invisible*, through an habitual sense of the high honour of belonging to Him, and a constantly accompanying and humbling sense of the sufferings of Him under whom we claim this honour, and the glad and grateful use of the means of exciting in us these apparently opposite sensations, is to fill ourselves with grace by prayer, to draw down those supplies of the Spirit of Jesus Christ which may stir us to go about doing good as He did, and to carry into *use and practice* what we have received. If we do not find God nigh unto us this way, it would be a contradiction to all the promises He has given us and all the precepts He has set before us, and one might well conclude that there is no such thing as *walking with Him.*

'Among other helps it might have been added that an habitual remembrance of our mortality, and that we are *dying creatures,* would keep us as it were in the presence of God, and that to have *Him* in our hearts and thoughts it is of course necessary to exclude as much as possible all such vain and worldly things, and all such vain and worldly persons as His Spirit would be offended with. And I conclude, as I began, that none of these things, or any other, can be of use to those who do not know how to use *Jesus Christ,* without whom it is not only folly but impious infidelity to think of approaching God, or praying to Him, or knowing anything of Him, or doing anything for Him.

'And now, my dear sir, if you do not find what I have said at all satisfactory, be assured you need not fear to own it, or to point out anything you think wrong or unintelligible in it. I fear it may bear evidence of the suffering state in which my head has long been, owing to the continuance of wet unsettled weather; and I have been obliged to write it at intervals, having had abundance of other writing to perform.

'I am, &c. &c.'

Little remains to be added to what has been already stated in this brief notice. Pursuing steadfastly the path of Christian duty from year to year, Mr. Whalley's life, for the sixteen years during which he occupied Chelwood, was a life of active charity, self-denial, and devotion. Ill health was his principal trial. Yet his life was a most happy one; satisfied to dwell in the seclusion of his humble parsonage, and intent on doing good to the utmost of his ability, he walked with God in the spirit of that perfect love which casteth out fear. The Bible was his constant and almost his sole study, and its great truths the bread of life to his soul. He had no wishes for preferment, no feelings of ambition—he sought not distinction or notice. Shunning, rather than seeking observation, he calmly pursued the noiseless tenor of his way; but when he did come in contact with the world, no one knew better what befitted a Christian and a gentleman. Though he meddled not in politics, he was a loyal subject and a sound Protestant, and jealous of every opinion tending to underrate the inspiration and authority of the Holy Scriptures.

He died in perfect peace, and with hopes full of immortality, on November 16, 1816, in the sixty-eighth year of his age.

POSTSCRIPT.

WILLIAM WILBERFORCE.

Soon after the death of the distinguished statesman and philanthropist William Wilberforce, a public meeting was held August 22nd, 1833, at which the Lord Chancellor Brougham presided, at which it was resolved:—

'That a subscription be opened for the purpose of doing honour to the memory of that distinguished person; first, by the erection of a monument, and secondly, if means be supplied, by such other methods as may be calculated to promote, in connection with the name of Wilberforce, the glory of God and the good of mankind.'

The fund raised for this purpose was invested in Government securities: the greater part of such securities were afterwards sold to defray the expenses of the monument erected in Westminster Abbey. Some time elapsing before the accounts connected with the monument could be made up, no steps were taken to decide upon the appropriation of the balance remaining in the public funds, and the matter passed out of mind. The treasurers of the fund, in whose names the money was invested, were Sir R. H. Inglis, Bart., and Sir T. F. Buxton, Bart.

After the death of the surviving treasurer, Sir R. H. Inglis, it was discovered that a balance of the Wilberforce Memorial Fund of 2,000*l*. 3 per cent. Consolidated Bank Annuities had stood in the names of the treasurers so long without the dividends being claimed, that at last the principal sum and the unclaimed dividends had been transferred, under the provisions of the Act, to the Commissioners for the reduction of the National Debt.

Upon the discovery of the oversight, Lord Brougham, and the remaining members of the original committee of the fund, who were called together for this purpose, petitioned the Court of Chancery for the appropriation of the money to its original object of a Wilberforce Memorial; and various suggestions having been considered, a scheme was submitted to the Court of Chancery, comprising the following points:—

'I. That it is desirable to appropriate a sum not exceeding 2,000*l*. of the fund in the erection of a building in Sierra Leone, to comprise—

1. A library of useful books, and reading-room.
2. Lecture rooms for lectures on science, literature, and religion.
3. A museum of practical industry.
4. Offices which may be used for a savings bank, in the hope that some well-digested and authorised plan for that essential object in the well-doing of the colony, may, ere long, be established in the colony—or such offices to be used for any other benevolent or useful purposes under the sanction of the trustees.

The sum of 400*l*. to be retained in this country, as a reserve fund for special purposes; and the remainder of the fund, after payment of all expenses incurred, and to be incurred in transaction of the business, to be devoted to the purchase of a library, philosophical instruments, and such other things as may be desired by the *ex-officio* trustees after mentioned.'

' II. That there is every reason to hope that the gift thus purposed to be made will be cordially welcomed in the colony, and to rely on a sufficient annual fund being raised there for the support of the buildings and institution, and other current expenses.

' III. That it be proposed that the Governor of Sierra Leone, the Bishop of Sierra Leone, the Chief Justice of Sierra Leone—respectively for the time being— and two other gentlemen to be nominated by the Church Missionary Society, be the trustees of the institution : and that the managing committee should consist of the trustees, three local subscribers to the funds of the institution to be chosen by the trustees, and six others of the local subscribers to be chosen by the whole body of the subscribers.'

The Court of Chancery approved of this scheme, and ordered (30th July, 1860) that the principal and interest of the Wilberforce Memorial Fund should ' be paid to the treasurer of the Church Missionary Society for Africa and the East, to the intent that the same may be applied by the said Society in carrying out the proposed scheme.'

The following regulations for carrying into effect the decree of Chancery have received the sanction of Lord Brougham and the members of the original committee, as well as the committee of the Church Missionary Society :—

1. *Erection of the Building and Ownership of the Property.*

The building is to be erected under the direction of the trustees. The property of the building, of the library, philosophical instruments, museum, and whatever else belongs to the institution, is to be vested in the trustees for the time being. All repairs or alterations in the fabric to be made only by their authority.

2. *Appointment of two Trustees by the Church Missionary Society.*

The two trustees to be appointed by the Church Missionary Society shall be appointed and removable from their office by a resolution of the Church Missionary Society. A copy of such resolution, certified by the signature of the majority of the secretaries of the society for the time being, shall be a sufficient appointment or removal from the office, to take effect upon the presentation of such resolution to the governor of the colony, or in case of his absence to the next senior resident trustee, in the order in which they are nominated by the decree in Chancery.

3. *The Library.*

As the library is to be provided in the first instance by the Church Missionary Society, so all future accessions to the books must be sanctioned by the Church Missionary Society, or by a majority of the existing trustees. The books in the library shall be annually compared with the catalogue, and the catalogue be revised accordingly, and such revised catalogue shall be sent over to the society each year.

4. *Committee of Management.*

The committee of management, when duly elected, shall draw up rules for the library, for the museum, and for lectures, and submit the same for the sanction of the trustees. The same committee shall have the appointment and removal of all officers or servants of the institution, and the regulation of their duties and their salaries, subject to an appeal to the trustees.

5. *Reserve Fund.*

The sum of 400*l.* to be reserved by the Church Missionary Society 'for special purposes,' and such portion as may remain after the payment of 2,000*l.* for the building, will be placed at interest until the committee is able to judge, after sufficient experience of the actual working of the institution, what department of the work may most require pecuniary aid, for the full carrying out of the decree of Chancery.

6. *Annual Reports.*

Annual reports of the institution shall be drawn up and printed for the use of subscribers on the spot, and twenty-five copies of the same shall be transmitted to the Church Missionary Society.

A very eligible site has been appropriated for the proposed institution in Freetown by Colonel Hill, the governor; and a plan and estimates for the proposed building have been prepared by the colonial surveyor and sent over to this country, and submitted to the approval of Lord Brougham and of the remaining members of the original committee. As soon as the society is assured that there is a sufficient number of persons in Sierra Leone prepared to form a managing committee, and to raise funds for the support of the institution, the building will be commenced. Donations of books, of a valuable and approved character, to be added to the proposed library, will be thankfully received at the Church Missionary House.

F. MAUDE, R.N.
Chairman.

Committee, Church Missionary House: *June,* 1862.

August 2, 1864.

We are happy to say the scheme has been matured under the advice of Lord Brougham and Sir Thomas Acland, and the first stone of the building was laid a few weeks ago by the Governor of Sierra Leone.

INDEX.

ACLAND, Sir Thomas, Bart., 70, 85, 184, 188, 282
Alexander I., Emperor of Russia, in London, 53, 55, 58
Arden, Mr. Pepper, 100
Artist, the young, and Benjamin West, 15

BABINGTON, Mr., 17, 158, 164, 186
Bank of England, stoppage of payment of the, 192
Bankes, Mr., 144, 145, 146, 171, 202
Barham, Lord, 242, 243
Barham Court, 242, 243
Barley Wood, 183, 268, 269
Barrington, Dr. Shute, Bishop of Durham, 282
Barry, Colonel, 93
Bath, Lord, 77
Battlefield, 291
Baxter, Richard, his works, 7
Beckwith, Major, at the Bristol riots, 232
Bellevue, Mr. Latouche's seat, 19, 135
Beresford, John Claudius, 192
Bible Society, British and Foreign, 1, 83, 97
Blaise Castle, 43, 90, 93, 225, 273
Blucher, Prince, 56
Boodle's Club, 205

Bourdaloue, anecdote of, 171
Bowdler, Mr., his death, 65
Boyle, Miss Kate, 21
Brereton, Colonel, at the Bristol riots, 232
Bristol, riots in, 232
Brockley Combe, 271
Brookes's Club, 205
Brougham, Mr. (now Lord), 56, 182, 249
Buchanan, Dr., his death, 66
Burgess, Dr., Bishop of St. David's, 281
Burke, Right Hon. Edmund, 94, 107, 141, 152, 180, 192, 278
Buxton, Sir Thomas Fowell, Bart., 123, 127

CAMBRIDGE University in the last century, 199
Camden, Pratt, first Lord, 130
Canning, Right Hon. G., 153, 181
Canova, 11
Caractacus, 291
'Cardiphonia,' 218
Carey, Rev. Dr., 177
Carlisle, Dean of, 89
Carrick, Dr., 285
Carter, Mrs., translator of Epictetus, 277
Castlereagh, Lord, 50, 152, 165
Catherine, Empress of Russia, 179

INDEX.

Cecil, Rev. Mr., 9
Chalmers, Rev. Dr., 108, 124
Chambers, Dr., 158, 159
Chatham, first Earl of, 154, 169
Chateaubriand, 62
Children, delights in, 115
Chillingworth's sermons, 131
Cholera of 1832, 237
Christian Knowledge Society, 131
Christophe, King of Hayti, 102
Cicero, his 'De Oratore,' 46
Clare, Lord, 193
Clarkson, Mr., 138
Clubs in the last century, 205
Coalbrookdale, 291
Coalition ministry, 203
Colthurst, Dr., 172, 173
Commons, House of, late hours in the, 100
Consalvi, Cardinal, 71, 75
Cooper, Rev. Mr., 290
Cottages in England, 169
Cowper, William, his poetry, 137, 143, 194
Crewe, Mrs., 108
Cuvier, Baron, 194

DALY, Rev. Robt., now Bishop of Cashel, 21
Davis, Mr. Hart, 9, 68, 190, 245
Death on the Pale Horse, West's picture of, 14
Derby, Earl of, 248
Devon, scenery of the north of, 69
Doddridge's 'Rise and Progress of Religion,' 206
Dolbyn, Sir W., 140
Douglas, the Hon. Frederick, 95
Dundas, Mr., 152
Dunn, Rev. James, 20, 21
Dunning, Mr., 174

EAST India Company, charter of, of 1813, 25

Eldon, John, first Lord, 97, 174, 183
Elwes, anecdotes of, 171
Elymas the sorcerer, 10

FAILAND, 272
Fitzwilliam, Earl, 186
Fox, Right Hon. C. J., 141, 146, 167. His India Bill, 203
France, restoration of the Bourbons in, 50. And the slave trade, 63
Freemasons' Hall, 2
Freemasons' Tavern, 56
French character, 170
Frenchay, 42
Fry, Mrs., at Newgate, 89

GARRICK, David, 10, 276
George, Prince Regent, 62
George III., 183
Giraldus Cambrensis, on the slave trade in Ireland, 91
Gisborne, Rev. T., 200, 201
Glenbervie, Lady, 94
Glendower's oak, 291
Gloucester, Duke of, 47
Gobble's Combe, 272
Godstone church and vicarage, 121
Goostree's house, in Pall Mall, 144
Gordon, Lord George, and General Murray, 102
Gordon, Mr., 138
Gordon, Duchess of, 169
Grant, Mr. Charles, 17, 128, 129
Grattan, Mr., 20
Gray, his ode to Cambridge University, 67, 68
Grenville, Lord, 154-156
Grey, Earl, 56, 57, 182. His Reform Bill, 230
Gurney, John Joseph, 123, 246

INDEX. 331

HALL, Rev. Robert, of Leicester, 98, 99. His sermons, 243
Hammer, M., 84
Harding, George, 174
Harrach, Count, 84
Harrowby, Lord, 47, 144, 155
Hayti, the 'Red Book' of, 61
Heber, Bishop, 131
Hensman, Rev. John, 285, 288
Hoare, Rev. Charles, Archdeacon of Winchester, 121
Holland, Richard Lord, 56
Holt, Chief Justice, 22, 23
Holwood, 139, 156
Hornby, Mrs., of Winwick, 21
Horne, Bishop, 133
Howard, John, 85
Hull, address from, to Mr. Wilberforce, 24. The town in 1745, 97. In the latter part of the last century, 199
Humboldt, Baron Von, 62

INDIA, Christian missions in, 26
Inglis, Sir Robert, 111, 256, 289
Interlaken, 207
Ireland, ancient civilisation of, 90. Rebellion of 1796 in, 192
Jebb, Rev. John, Bishop of Limerick, 22, 133
Johnson, Dr. Samuel, 101, 168, 274, 275, 277, 278
Judge, the, and 'the deaf culprit,' 101

KAYE, Dr., Bishop of Bristol, 131
Kensington Gore, 4, 5, 53
Killerton Park, 70
Knox, Mr. Alexander, 20, 22, 239

LADY, death of a Christian, 87
Lampeter College, 187

Lansdowne, Marquis and Marchioness of, 47, 56
Latouche, Peter, Esq., 19, 135
Lawrence, Sir Thomas, his portrait of Mr. Wilberforce, 256, 257
Leisure, retrospect of, 113
Limonade, Count, 102
Linton, visit to, 69
Liverpool, Lord, 224
Locke, John, his birth-place, 272
London in 1814, 53, 58. Thieves of, 175

MACAULAY, Mr. (afterwards Lord), 189. His visits to Barley Wood, 280, 281
Macaulay, Mr. Zachary, 240
Mackintosh, Sir James, 47, 99
Mackworth, Major (afterwards Sir Digby), 233
Malvern, 290
Manchester, Duke of, Ambassador to the court of Paris, 150
Mansfield, Lord, 174
Marden Park, 112, 120
Marie Antoinette, Queen, 148, 149, 170
Marsh, Mr., 30
Matlock, 290
'Mendip Annals,' Miss Patty More's, 269
Middleton, Lady, 242
Milner, Dean, 58, 59, 206
Milton, Johnson's sayings of, 278
Missions, Christian, in India, 26. Mr. Wilberforce's speech on, 31
Monboddo, Lord, and Hannah More, 277
Montgomery's Hymns, 225
Montrose, Duke of, 98
More, Hannah, 3, 29, 122, 235, 240, 242, 245. Brief recollections of, 268. Her sisters, 269. Her personal appearance, 270. Her first interview with Dr. Johnson, 270. Her friend-

ship with Mr. Garrick, 276.
Her refusal of Lord Monboddo,
277. Visitors at her house,
279. Her benevolence, 279.
Her guest, Tom (afterwards
Lord) Macaulay, 280. Her
intimate friends and correspondents, 281. Her portrait by
Pickersgill, 282. Loss of her
sisters, 283. Her removal to
Clifton, 285. Her last moments, 287. Her death, 289.
Her letter to a family in deep
affliction, 292
More, Miss Betty, 270
More, Miss Patty, 192. Her
'Mendip Annals.' 269. Her
death-bed, 283. Her character, 284
More, Miss Sally, 269
Murderer, 176
Murray, General, and Lord
George Gordon, 102

NAPOLEON I. and Benjamin
West, 14, 15. The Russian
campaign, 47
Newgate, Mrs. Fry at, 89
Newton, Rev. John, 107, 209, 218
Noailles, Marquis de, 150
Nollekens, his statue of Pitt, 11
North, Lord, 93
North, Frederick (afterwards Earl
of Guildford), 78, 95

OLDENBURGH, Duchess of, 58
Oxford, 67

PALACE Yard, Mr. Wilberforce's residence in, 1, 4, 97
Pelet, Madame, 187
Perceval, Right Hon. Mr., assassination of, 17. His personal character, 18

Peltier, his 'Ambigu,' 61
Pitt, Right Hon. W., 11, 45, 139.
141, 143, 144 *et seq.*, 201,
202, 207, 210, 211
Pitt, Mr. Morton, 175
Pius VII., 79, 81
Porteus, Dr. Beilby, Bishop of
London, 107, 281
Portugal, slave trade of, 72
Potemkin, Prince, 179
Prussia, King of, in London,
1814, 54, 55

QUEENSBERRY, Duke of, at Richmond, 169

RAIKES, Rev. Henry, his sermons, 226
Raphael, cartoons of, 10
Raynal, the Abbé, and Dr. Johnson, 277
Reform Bill, 230, 231
Rendlesham, Lord, 147
Revolution, French, 152
Reynolds, Richard, 43, 291
Rheims, Messrs. Pitt, Wilberforce,
and Eliot at, 147, 148
Richmond, George, his portrait
of Mr. Wilberforce, 260
Rokeby, 187
Russia, Napoleon's disastrous
campaign in, 47
Ryder, Mr. (afterwards Lord
Harrowby), 130, 144
Ryder, Dr., Bishop of Lichfield,
68, 237, 238, 290

'SCOTSMAN' newspaper, 121
Scott, Sir Walter, 111, 112
Scott, Sir John (afterwards Lord
Eldon), 174. *See* Eldon
Sedgwick, Prof., 142
Selwyn, George, 169, 205
Sharp, Granville, 138
Shelburne, Lord, 145, 146, 202

Sheridan, Right Hon. R. B., his speeches, 46, 166
Sierra Leone, testimonial to Mr. Wilberforce at, 323.
Sismondi, M., on the slave trade, 60. His other works, 61, 62
Slave trade, 51, 55, 56, 73, 138. Partly prohibited by France, 63. Spanish and Portuguese, 72, 79. Slave trade in Ireland in former ages, 91. Abolition Bill passed, 246
Slave ships, 73
Smith, Mr. Abel, 197
South's sermons, 131
Spain, slave trade of, 72, 77. Treaty with, 83
Staël, Baron de, 120
Staël, Madame de, 46, 62
Staffa and Iona, visit to, 162, 164
Stanley, Lord (now Earl of Derby), 248
Stephen, Mr., 17, 19, 236
Stephen, Mrs., her death, 76
St. Helen's, Lord, 179
Sunday at Kensington Gore, 6
Switzerland, scenery of, 207

TALLEYRAND, Prince, 59, 61
Teignmouth, Lord, 96, 97
Tierney, Mr., 56
Thelluson, Mr. (afterwards Lord Rendlesham), 147
Thieves of London, 175
Thornton, Mr. Henry, 1. His death, 65
Tintern Abbey, 68
Trevelyan, Rev. Walter, 268

VATICAN, the, 80

WARREN, Dr., 90
Water's-meet, 69

Way, Sir Gregory, 167
Wellington, Arthur, Duke of, 47, 60, 61, 167
Wesley, Rev. John, 22
West, Benjamin, 9–16
Whalley, Rev. R. C., 46. Sketch of his life and character, 294. His birth and parentage, 294. His early views of religion, 295. Hannah More's description of him, 295. Peruses 'Butler's Analogy,' 296. His marriage, 297. Takes holy orders, 297. Loses his wife, 298. His account of himself, 298. Renovation of his heart, 302. His sermons, 304. His sacramental prayer, 305. Goes to Chelwood, 306. His letters, 307, 311, 312, 316. Last years of his life, 321. His death, 322.
White's Club, 205
Wilberforce, William, Esq., M.P., his personal appearance, 2. His domestic habits, 4. His guests, 9. Resigns his seat for Yorkshire, 24. Becomes member for Bramber, 24. His speech on Indian missions, 31. His hints as to public speaking, 42. His charity, 49. His speech on France and the Slave Trade, 51. At Freemasons' Hall, 56. His illness in 1788, 90. His residence in Palace Yard, 97. His 'Practical View of Christianity,' 103. Removes to Marden Park, 112. Death of his eldest daughter, 115. His appeal to his countrymen on Negro Slavery, 126. His thoughts as to retiring from public life, 131. Why he was first induced to take up the Slave Trade question, 138. Cowper's sonnet to him, 143.

In France with Mr. Pitt, 147. His retirement from Parliament, 158. His fondness for music, 168. His mental powers at seventy, 191. His details of some of the leading particulars of his life, 196 et seq. Loss of his surviving daughter, 234. His last visit to the author, 237. His speech at Maidstone on slavery, 241. His last illness, 246-248. His death, 249. His funeral, 251. Public manifestations of sorrow, 252 et seq. His character and domestic habits, 255. The Rev. Dr. Hawkins's letter respecting him, 265. His remarks on a letter of the Rev. Richard Whalley, 308. The testimonial to him at Sierra Leone, 323.

Wilberforce, Mrs., 117
Wilson, Rev. Daniel, 89
Windham, Mr., 45, 101, 123, 143
Witherspoon's 'Practical Essay on Regeneration,' 213
Wrington, 272
Wye, scenery of the, 68

YORK, the Wilberforce memorial in, 253
Young, Arthur, 9

LONDON
PRINTED BY SPOTTISWOODE AND CO.
NEW-STREET SQUARE

By the same Author.

LIFE OF MICHAEL ANGELO BUONARROTI;
WITH
TRANSLATIONS OF MANY OF HIS POEMS AND LETTERS.

In Two Volumes.

SECOND EDITION.

www.ingramcontent.com/pod-product-compliance
Lightning Source LLC
Chambersburg PA
CBHW032357230426
43672CB00007B/730